SAS
Programming
The One-Day Course

NEIL SPENCER

CHAPMAN & HALL/CRC

A CRC Press Company
Boca Raton   London   New York   Washington, D.C.

Library of Congress Cataloging-in-Publication Data

Spencer, Neil.
      SAS programming : the one-day course / Neil Spencer
              p.   cm.
      Includes index.
      ISBN 1-58488-409-6
      1. SAS (Computer file) 2. Statistics—Data processing. I. Title.

QA276.4.S6455 2003
519.5′.0285—dc21   2003055157
                                                                  CIP

Visit the CRC PressWeb site at www.crcpress.com

© 2004 by Chapman & Hall/CRC

No claim to original U.S. Government works
International Standard Book Number 1-58488-409-6
Library of Congress Card Number 2003055157
Printed in the United States of America 1 2 3 4 5 6 7 8 9 0
Printed on acid-free paper

Dedication

for

Catherine, Laura, Julia, and Helen

Preface

This book has been created by the author as material for a one-day course in SAS programming. The fact that it was developed for a training course is reflected in the concise nature of the presentation. It is not intended to be a general reference book for programming in SAS but is aimed at researchers and students who want to learn the basics of data management, summarising data, and graphics. Once readers have mastered the topics covered in this book, they will be well placed to learn further aspects of SAS programming for themselves. Information on these further aspects can be obtained from the SAS help facilities, relevant manuals, and the SAS Institute Web pages (http://www.sas.com).

Each chapter apart from Chapter 1 contains tasks so that the reader can practice the programming presented in that chapter. The data files needed to carry out the tasks can be found online via: http://www.crcpress.com/e_products/downloads/. The suggested solution programs shown in Appendix C can also be found in electronic form on this website. Also contained at this location are electronic versions of the programs and macros used in the text and the data they access.

The reader is, of course, not obliged to work through the book in just one day, but the material is arranged in such a way that programs and tasks sometimes use aspects of programming or datasets created in previous chapters. It is thus suggested that the reader tackle the topics in the order presented.

SAS Version 8.2 was used by the author in preparing this book, and at the time of writing, SAS was rolling out Version 9. As one would expect, literature describing Version 9 states that programs written for earlier versions of SAS will work under the new version.

Dr. Neil H. Spencer
Principal Lecturer in Statistics, University of Hertfordshire
N.H.Spencer@herts.ac.uk

Author

Neil Spencer is a principal lecturer in statistics at the University of Hertfordshire and head of its Statistical Services and Consultancy Unit. He undertook his undergraduate studies at the University of Reading (where he first encountered SAS), followed by postgraduate study at the University of Southampton and doctoral study at the University of Lancaster. At the University of Hertfordshire, he teaches various courses, including SAS programming to undergraduate students and as a short course through the University's Statistical Services and Consultancy Unit. His research is primarily based in the social sciences and includes multilevel modelling of educational data, statistical computing, and multivariate statistics.

Table of Contents

1 Introducing SAS

SAS is a powerful computer package used for the analysis of data. It is made up of various modules that contain PROCEDURES to undertake specific tasks. The BASE module contains the basics of the SAS system, including data entry and manipulation and simple statistical operations. Purchasers of SAS can then decide what additional modules they require. Examples include the SAS/GRAPH module for graphics, the SAS/IML module (*I*nteractive *M*atrix *L*anguage), and the SAS/STAT module for more advanced statistical operations. There is also a SAS/ASSIST module that allows the user to interact with the system directly, in a similar manner as to many other statistical software packages. It does not, however, use Microsoft® Windows® dialogue boxes that users might expect, and it is much less user friendly than other packages in this regard.

Unless SAS/ASSIST or some other interactive system is being used, the package is operated by the user writing a program and then submitting it to the system. The system then processes the statements and produces output, as appropriate, and a log file, giving details of what operations were carried out.

So why does SAS not have a user-friendly interface? Why does it expect its users to learn how to write programs? The answers lie in the nature of research projects: they are typically undertaken by a group of researchers and are ongoing for a significant amount of time. This means that a record must be kept of exactly what data manipulations, analyses, etc., were accomplished. Keeping programs of statements, which can contain comments on what the programs do, what revisions were made, dates on which the programs were created, and when revisions were made, is an effective tool. These programs can also contain statements that carry out many analyses and data manipulations within the same program. When revisions are made to these more complex programs, the resulting program can be simply submitted to the system, and all the analyses and data manipulations can be carried out together. To do the same using dialogue boxes may necessitate the use of many boxes and take a significant amount of operator time.

SAS is not unique in offering this ability to write programs in a command language. Many packages that use dialogue boxes simply convert the dialogue box information into their own command language for processing. Some keep this conversion hidden from the user, some show it explicitly, and others offer the option of displaying the converted commands.

The capabilities of the SAS system are immense. With this book, the author does not expect readers to become expert programmers in one day but does aim to develop skills that will enable readers to use SAS to a certain level and, perhaps more importantly, put them in a position to further develop programming skills in the particular areas of application that are relevant to their work.

In the programs in this book, words that are part of the SAS programming language are given in uppercase, and words that are defined by the user (e.g., variable names and dataset names) are given in lowercase. SAS, however, is not case sensitive, and the distinction shown here is for clarity only.

1.1   OPERATING SAS

How the user interacts with SAS depends to a large extent on the operating system that is being used. It is not the intention of this book to document in detail the vagaries of using SAS with every possible operating system — its aim is to introduce the programming language. However, some features are common to all systems (in one form or another) and are discussed below. Throughout the rest of the book, efforts are made to avoid detail specific to any particular operating system. The only exception is when the author has judged that mention of features found in the widespread Microsoft Windows operating system is worthwhile (e.g., Section 1.2). When external files are accessed in programs, MS-DOS paths are used.

1.1.1   WRITING AND SUBMITTING A SAS PROGRAM

A SAS program is written in plain text, and any text editor or word processor that can save in plain text format can be used. Once complete, the program is submitted to the SAS software for processing.

For the sake of efficiency, it is a good idea to restrict individual programs to performing just a few tasks. Programs that undertake many different analyses and data manipulations can become unwieldy and difficult to understand.

1.1.2   SAS LOG

Once SAS has processed a submitted program, it displays the operations it carried out in a log file, which displays the submitted program in sections with "Notes," "Warnings," and "Errors." Notes are good, warnings are potentially worrisome, and errors tell the user why the program failed. As with any computer program, it is unusual for a SAS program to be written correctly the first time. An understanding of what the log window is telling the user about the program submitted is vital in debugging and is developed as a programmer gains experience.

1.1.3   SAS OUTPUT

If the submitted program creates any nongraphical viewable output (i.e., analysis results or tables), then it will be displayed by default in an output file. Graphical output will be produced in a format appropriate to the operating system being used. If the Output Delivery System (discussed in Section 8.1) is used, other files may result.

1.2   OPERATING SAS WITH MICROSOFT WINDOWS

In Version 8 of the Microsoft Windows version of SAS, there is an "Enhanced Program Editor" window, which is where the SAS program resides. When this window is active (has a coloured bar at the top rather than a grey bar), the menu bar at the top of the screen is particular to the Program Editor, and the toolbar buttons will also operate in the window. There are facilities for opening, closing, and saving programs as well as cutting, copying, and pasting. It operates simply as a text editor. An advantage of using the SAS Enhanced Program Editor over other plain text editors is that it automatically colour codes different parts of the program and displays horizontal lines dividing separate parts of the program. This is a valuable aid when writing a program, as when errors in programming syntax are made, the user does not see the expected colour patterns and thus knows that something is wrong. On the toolbar, there is a button that is selected to run the program. The program can also be run via the menu bar: "Run", "Submit".

1.2.1 OTHER WINDOWS

The "Output" window is where the results of the program are shown by default. There is also an accompanying "Results" window that lists the output produced in folder-like forms. SAS organises its output into "pages". The number of lines that can be fitted onto a page and the maximum length of these lines can be customised by using "Tools", "Options", "System…" and then, within the "Log and procedure output control" folder, selecting "SAS log and procedure output". The option that the user wishes to change is then selected. When the Output window is printed directly from SAS, each new page in SAS requires a fresh sheet of paper. It is frequently the case that a page of SAS output contains only a few lines.

The "Log" window is important when there are errors in the program that was written. It displays the submitted program in sections with colour-coded "Notes", "Warnings", and "Errors".

If graphs are produced by SAS as a result of a program running, they open in separate windows.

1.2.2 CLEARING TEXT FROM WINDOWS

When an SAS program is run, edited, and rerun successively until a correctly working program is obtained, a lot of redundant text will accumulate in the Log window and potentially in the Output window.

Thus, it is often a good idea to clear all the text from a window before submitting the program. The text that then appears in the Log window must be relevant to the program just submitted, and debugging can take place if necessary. It can also be confusing to have irrelevant text in the Output window, so once a correctly operating program is obtained, it is a good idea to run it again after clearing all previous text from the Output window.

2 Reading Data into SAS

Unlike many statistical packages, SAS can operate with several datasets at one time. Some of the flexibility that this offers will be investigated later, but the first step in any analysis has to be to access the data. The issue of permanent SAS datasets will also be addressed later, but the first issue addressed is that of how data can be taken into SAS from a plain text file as part of a SAS program that may also contain statements to carry out analyses on the data. Using PROC IMPORT, SAS can also import data from a delimited plain text file and data that were previously saved by several different software packages. The use of this procedure will also be examined here.

2.1 WHAT IS PLAIN TEXT?

Data are in plain text format when they are entered into a text editor or are saved as text from a word processor (e.g., "Text Only" form in Microsoft Word®) or a spreadsheet (e.g., "Formatted Text [Space delimited]" in Microsoft Excel®). A data file with TAB characters between columns of data can also be used but can cause confusion, because the TAB character counts as one character to SAS, even though it may look like it takes up several spaces in the text file.

2.2 TUMOUR NECROSIS FACTOR (TNF) DATA

Data from a study into the production of tumour necrosis factor (TNF) by cells in laboratory conditions are shown in Table 2.1. The study involved an investigation of two different stimulating factors, mycobacterium tuberculosis (MTB) and fixed activated T-cells (FAT), added individually and together. Cells from 11 donors were used.

TABLE 2.1
Tumour Necrosis Factor (TNF) Data

		No MTB					MTB		
FAT	Donor	TNF, Three Replicates			FAT	Donor	TNF, Three Replicates		
No	1	-0.01	-0.01	-0.13	No	1	-0.05	-0.09	-0.08
No	2	16.13	-9.62	-14.88	No	2	-9.41	-6.3	5.4
No	3	Missing	-0.3	-0.95	No	3	-10.13	-16.48	-14.79
No	4	3.63	47.5	55.2	No	4	8.75	134.9	203.7
No	5	-3.21	-5.64	-5.32	No	5	612.3	898.2	854.2
No	6	16.26	52.21	17.93	No	6	2034	2743	2772
No	7	-12.74	-5.23	-4.06	No	7	978.5	1137	850
No	8	-4.67	20.1	110	No	8	279.3	124.8	222.1
No	9	-5.4	20	10.3	No	9	688.1	530.9	720.2
No	10	-10.94	-5.26	-2.73	No	10	908.8	811.9	746.4
No	11	-4.19	-11.83	-6.29	No	11	439.3	960.9	593.3
Yes	1	88.16	97.58	66.27	Yes	1	709.3	874.3	630
Yes	2	196.5	114.1	134.2	Yes	2	4541	4106	4223

Yes	3	6.02	1.19	3.38	Yes	3	391	194	254
Yes	4	935.4	1011	951.2	Yes	4	2913	3632	3417
Yes	5	606	592.7	608.4	Yes	5	3801	3112	3681
Yes	6	1457	1349	1625	Yes	6	10150	9410	9243
Yes	7	139.7	399.5	91.69	Yes	7	6736	6323	5117
Yes	8	196.7	270.8	160.7	Yes	8	1454	2250	1092
Yes	9	135.2	221.5	268	Yes	9	857.2	1339	1945
Yes	10	-14.47	79.62	304.1	Yes	10	Missing	739.9	4379
Yes	11	516.3	585.9	562.6	Yes	11	6637	6909	6453

Source: Data are from Dr. Jan Davies, published in Bland, M., *An Introduction to Medical Statistics*, 3rd ed., Oxford University Press, Oxford, 2000. Reprinted by permission of Oxford University Press.

The TNF data shown in Table 2.1 are available in a plain text file "tnfdata.dat". Part of this file is shown below.

```
001 -0.01  -0.01  -0.13
002 16.13  -9.62  -14.8
003         -0.3  -0.95
004   3.63   47.5  55.2
.
.
.
118  1454   2250  1092
119 857.2   1339  1945
1110        739.9 4379
1111 6637   6909  6453
```

Column 1 contains a 0 for No MTB and a 1 for MTB.
Column 2 contains a 0 for FAT = No and a 1 for FAT = Yes.
Columns 3 to 4 contain the donor number.
Columns 5 to 10 contain the data for the first replicate.
Columns 11 to 16 contain the data for the second replicate.
Columns 17 to 22 contain the data for the third replicate.

Program 2.2.1 may be used to read this data file. The PROC PRINT procedure is used to show the data.

Program 2.2.1

```
*Author: Neil Spencer;
*Creation Date: ??/??/??;
*Revision 1 Date: ??/??/??;
*Purpose: Read TNF data;
*Purpose: Program 2.2.1;
Comments start with * and end with a semicolon.


DATA tnfdata;
The dataset is given the name "tnfdata".
```

```
INFILE 'a:\data\tnfdata.dat' TRUNCOVER;
```
The TRUNCOVER option is included, because some of the lines in the dataset have the variable "rep3" ending before the stated column 22.

```
INPUT mtb 1 fat 2 donor 3-4 rep1 5-10
    rep2 11-16 rep3 17-22;
```
The INFILE statement tells SAS where the dataset file is located, and the INPUT statement tells SAS what the variables are called and in which columns they may be found.

```
PROC PRINT DATA = tnfdata;
```
The PROC PRINT displays the "tnfdata" dataset.

```
RUN;
```
The final RUN statement tells SAS that the program has finished and, having been read in, should be executed.

SAS Output for Program 2.2.1

Obs	mtb	fat	donor	rep1	rep2	rep3
1	0	0	1	-0.01	-0.01	-0.13
2	0	0	2	16.13	-9.62	-14.80
3	0	0	3	.	-0.30	-0.95
4	0	0	4	3.63	47.50	55.20
5	0	0	5	-3.21	-5.64	-5.32
6	0	0	6	16.26	52.21	17.93
7	0	0	7	-12.74	-5.23	-4.06
8	0	0	8	-4.67	20.10	110.00
9	0	0	9	-5.40	20.00	10.30
10	0	0	10	-10.94	-5.26	-2.73
11	0	0	11	-4.19	-11.83	-6.29
12	0	1	1	88.16	97.58	66.27

13	0	1	2	196.50	114.10	134.20
14	0	1	3	6.02	1.19	3.38
15	0	1	4	935.40	1011.00	951.20
16	0	1	5	606.00	592.70	608.40
17	0	1	6	1457.00	1349.00	1625.00
18	0	1	7	139.70	399.50	91.69
19	0	1	8	196.70	270.80	160.70
20	0	1	9	135.20	221.50	268.00
21	0	1	10	−14.47	79.62	304.10
22	0	1	11	516.30	585.90	562.60
23	1	0	1	−0.05	−0.09	−0.08
24	1	0	2	−9.41	−6.30	5.40
25	1	0	3	−10.13	−16.48	−14.70
26	1	0	4	8.75	134.90	203.70
27	1	0	5	612.30	898.20	854.20
28	1	0	6	2034.00	2743.00	2772.00
29	1	0	7	978.50	1137.00	850.00
30	1	0	8	279.30	124.80	222.10
31	1	0	9	688.10	530.90	720.20
32	1	0	10	908.80	811.90	746.40
33	1	0	11	439.30	960.90	593.30
34	1	1	1	709.30	874.30	630.00
35	1	1	2	4541.00	4106.00	4223.00
36	1	1	3	391.00	194.00	254.00
37	1	1	4	2913.00	3632.00	3417.00
38	1	1	5	3801.00	3112.00	3681.00
39	1	1	6	10150.00	9410.00	9243.00
40	1	1	7	6736.00	6323.00	5117.00
41	1	1	8	1454.00	2250.00	1092.00
42	1	1	9	857.20	1339.00	1945.00
43	1	1	10	.	739.90	4379.00
44	1	1	11	6637.00	6909.00	6453.00

Additional Notes for Program 2.2.1

Semicolons — Note that in Program 2.2.1, every statement ends with a semicolon. If there is no semicolon at the end of a line in a program, SAS simply continues reading the same statement from the next line until it comes to a semicolon. One of the most common mistakes in SAS programming is forgetting the semicolon. Unfortunately, when this happens, the next statement is read as a continuation of the current statement, and SAS reports a syntax error relating to the statement. It is up to the user to notice that the error is not due to bad syntax but is due to a missed semicolon.

Comments in programs — Program 2.2.1 started with a series of comments (each line beginning with an "*" and ending with a ";"). These can be used to keep a record of, for example, who wrote the program, when it was created, when it was modified, and its purpose. For an individual writing many SAS programs, it can help keep track of the work that was done, and (perhaps more importantly) for a person working as part of a team, others can see what was done. An alternative way of writing a series of comments (a section of a program beginning with an "*" and ending with an "*/") can be seen in Program 2.3.1.

Dataset names — In Program 2.2.1, a dataset called "tnfdata" is created. The choice of name is up to the user and can contain letters, digits, and the "_" character, as long as the first character of the name is not a digit. The maximum length of the dataset name is 32 characters. It is a good idea to use a name that is meaningful to the analysis or project being undertaken.

Column numbers — In Program 2.2.1, SAS was told what columns each variable was occupying. If the plain text data file contained the variables in columns with spaces between the columns, these column numbers could have been omitted. However, in this case, if a blank space is used to denote a

missing value, the MISSOVER option is needed for the
INFILE statement.

Reading text variables — In Program 2.2.1, all of the variables
being read are numeric. Text variables are allowed in SAS,
but to read them in, the "$" symbol must be used immedi-
ately following the variable name. An example would be
`INPUT name $ 1-10 age 11-12;`.

Variable names — The variable names given to variables are
restricted to a maximum of 32 characters. As with dataset
names, they can contain letters, digits, and the "_" character,
as long as the first character of the name is not a digit.

Saving programs — When amendments are made to a SAS
program, it is advisable to save it. In the Microsoft® Win-
dows® environment, this can be done in the usual manner
for Microsoft Windows-based software via "File, Save/Save
As..." or the toolbar.

`INFILE 'a:\data\tnfdata.dat' TRUNCOVER;` — It
should be noted that SAS does not necessarily need the entire
path of the location of the file (the "a:\data\") in order to find
it. Depending on the operating system and how SAS was
installed, there will be default locations for files. However, it
is often a good idea to use full or near-full paths, as this reduces
the potential for confusion. It should also be noted that in recent
versions of the Microsoft Windows environment, it is often the
case that the default is for filename extensions (the ".dat" in
"a:\data\tnfdata.dat") not to be displayed if Microsoft Windows
recognises the extension. This happens if a program installed
on the computer uses the extension as an indicator of a particular sort of file. It is a good idea to turn off this feature so that
the full filename needed by SAS can be seen.

`PROC PRINT DATA = tnfdata;` — The only item contained
in Program 2.2.1 before the PROC PRINT statement is the
reading in of the data into a dataset called "tnfdata". Thus, if
the statement omitted "DATA = tnfdata", SAS would realise
that the PROC PRINT should be applied to the dataset "tnfdata". However, it is good practice to include "DATA = tnf-

data" in the PROC PRINT so that it is clear to other readers
of the program what data it is meant to be reading. Also, when
several datasets are being used in one program, including
"DATA = tnfdata" means there is no chance of the wrong
dataset being used by accident.

2.3 GIVING LABELS TO VARIABLES

In Program 2.2.1, variables were given names with a maximum of 32
characters, but because they cannot contain spaces, these names often
did not describe the variables well. Longer labels that are more descrip-
tive can be attached to the variables using the LABEL statement, as
shown in Program 2.3.1.

PROGRAM 2.3.1

```
/*
Author: Neil Spencer
Creation Date: ??/??/??
Revision 1 Date: ??/??/??
Purpose: Produce average TNF variable
Purpose: Program 2.3.1
*/
```
Note: Comments are now in a block starting with / and
ending with */. Semicolons are not required with this format
for comments.*

```
DATA tnfdata;
INFILE 'a:\data\tnfdata.dat' TRUNCOVER;
INPUT mtb 1 fat 2 donor 3-4 rep1 5-10
  rep2 11-16 rep3 17-22;
LABEL rep1 = '1st replicate';
LABEL rep2 = '2nd replicate';
LABEL rep3 = '3rd replicate';
LABEL mtb = 'MTB presence';
```

```
LABEL fat = 'FAT presence';
```
*Labels are given to some of the variables to make interpre-
tation of output easier.*

```
PROC PRINT DATA = tnfdata;
RUN;
```

SAS Output for Program 2.3.1

This is the same as the output for Program 2.2.1. Variable labels
are not used in PROC PRINT by default but would be used if
the LABEL option were added to the PROC PRINT statement.

2.4   CATEGORY LABELS

As well as variables having labels attached to them, discrete variables
can have labels attached to their categories using the PROC FORMAT
procedure. An example is provided in Program 2.4.1.

PROGRAM 2.4.1

Comments omitted

```
PROC FORMAT;
VALUE mtblab 0 = 'No MTB' 1 = 'MTB present';
VALUE fatlab 0 = 'No FAT' 1 = 'FAT present';
```
*PROC FORMAT sets up the labels for values of the categorical
variables.*

```
DATA tnfdata;
INFILE 'a:\data\tnfdata.dat' TRUNCOVER;
INPUT mtb 1 fat 2 donor 3-4 rep1 5-10
  rep2 11-16 rep3 17-22;
LABEL rep1 = '1st replicate';
LABEL rep2 = '2nd replicate';
```

```
LABEL rep3 = '3rd replicate';
LABEL mtb = 'MTB presence';
LABEL fat = 'FAT presence';

FORMAT mtb mtblab.; FORMAT fat fatlab.;
```
*The formats are applied to the categorical variables. Note the
full stop after "mtblab" and "fatlab."*

```
PROC PRINT DATA = tnfdata;
RUN;
```

SAS Output for Program 2.4.1

Obs	mtb	fat	donor	rep1	rep2	rep3
1	No MTB	No FAT	1	−0.01	−0.01	−0.13
2	No MTB	No FAT	2	16.13	−9.62	−14.80
3	No MTB	No FAT	3	.	−0.30	−0.95
4	No MTB	No FAT	4	3.63	47.50	55.20
5	No MTB	No FAT	5	−3.21	−5.64	−5.32
6	No MTB	No FAT	6	16.26	52.21	17.93
7	No MTB	No FAT	7	−12.74	−5.23	−4.06
8	No MTB	No FAT	8	−4.67	20.10	110.00
9	No MTB	No FAT	9	−5.40	20.00	10.30
10	No MTB	No FAT	10	−10.94	−5.26	−2.73
11	No MTB	No FAT	11	−4.19	−11.83	−6.29
12	No MTB	FAT present	1	88.16	97.58	66.27
13	No MTB	FAT present	2	196.50	114.10	134.20
14	No MTB	FAT present	3	6.02	1.19	3.38
15	No MTB	FAT present	4	935.40	1011.00	951.20
16	No MTB	FAT present	5	606.00	592.70	608.40

17	No MTB	FAT present	6	1457.00	1349.00	1625.00
18	No MTB	FAT present	7	139.70	399.50	91.69
19	No MTB	FAT present	8	196.70	270.80	160.70
20	No MTB	FAT present	9	135.20	221.50	268.00
21	No MTB	FAT present	10	−14.47	79.62	304.10
22	No MTB	FAT present	11	516.30	585.90	562.60
23	MTB present	No FAT	1	−0.05	−0.09	−0.08
24	MTB present	No FAT	2	−9.41	−6.30	5.40
25	MTB present	No FAT	3	−10.13	−16.48	−14.70
26	MTB present	No FAT	4	8.75	134.90	203.70
27	MTB present	No FAT	5	612.30	898.20	854.20
28	MTB present	No FAT	6	2034.00	2743.00	2772.00
29	MTB present	No FAT	7	978.50	1137.00	850.00
30	MTB present	No FAT	8	279.30	124.80	222.10
31	MTB present	No FAT	9	688.10	530.90	720.20
32	MTB present	No FAT	10	908.80	811.90	746.40
33	MTB present	No FAT	11	439.30	960.90	593.30

34	MTB present	FAT present	1	709.30	874.30	630.00
35	MTB present	FAT present	2	4541.00	4106.00	4223.00
36	MTB present	FAT present	3	391.00	194.00	254.00
37	MTB present	FAT present	4	2913.00	3632.00	3417.00
38	MTB present	FAT present	5	3801.00	3112.00	3681.00
39	MTB present	FAT present	6	10150.00	9410.00	9243.00
40	MTB present	FAT present	7	6736.00	6323.00	5117.00
41	MTB present	FAT present	8	1454.00	2250.00	1092.00
42	MTB present	FAT present	9	857.20	1339.00	1945.00
43	MTB present	FAT present	10	.	739.90	4379.00
44	MTB present	FAT present	11	6637.00	6909.00	6453.00

Additional Notes for Program 2.4.1

Positioning of PROC FORMAT — PROC FORMAT must be positioned before the data step. This is because SAS is being flexible and allowing the formats defined by the PROC FORMAT potentially to be applied to any live dataset.

FORMAT mtb mtblab.; FORMAT fat fatlab.; — Note that SAS allows two statements on one line here because it sees the first semicolon as being the end of a statement. What then follows must be the beginning of a new statement.

2.5   USING SIMPLE ARITHMETIC TO CREATE A NEW VARIABLE

Three replicates of TNF exist for each case in the dataset. From these, Program 2.5.1 creates a new variable that is the average of the three replicates.

PROGRAM 2.5.1

```
Comments omitted

PROC FORMAT;
VALUE mtblab 0 = 'No MTB' 1 = 'MTB present';
VALUE fatlab 0 = 'No FAT' 1 = 'FAT present';

DATA tnfdata;
INFILE 'a:\data\tnfdata.dat' TRUNCOVER;
INPUT mtb 1 fat 2 donor 3-4 rep1 5-10
   rep2 11-16 rep3 17-22;
LABEL rep1 = '1st replicate';
LABEL rep2 = '2nd replicate';
LABEL rep3 = '3rd replicate';
LABEL mtb = 'MTB presence';
LABEL fat = 'FAT presence';
FORMAT mtb mtblab.; FORMAT fat fatlab.;

tnfaver = (rep1+rep2+rep3)/3;
LABEL tnfaver = 'average TNF';
```
The new variable, "tnfaver" is created using simple arithmetic.

```
PROC PRINT DATA = tnfdata;
```
The PROC PRINT displays the updated "tnfdata" dataset.

```
RUN;
```

SAS Output for Program 2.5.1

Obs	mtb	fat	donor	rep1	rep2	rep3	tnfaver
1	No MTB	No FAT	1	-0.01	-0.01	-0.13	-0.05
2	No MTB	No FAT	2	16.13	-9.62	-14.80	-2.76
3	No MTB	No FAT	3	.	-0.30	-0.95	.
4	No MTB	No FAT	4	3.63	47.50	55.20	35.44
5	No MTB	No FAT	5	-3.21	-5.64	-5.32	-4.72
6	No MTB	No FAT	6	16.26	52.21	17.93	28.80
7	No MTB	No FAT	7	-12.74	-5.23	-4.06	-7.34
8	No MTB	No FAT	8	-4.67	20.10	110.00	41.81
9	No MTB	No FAT	9	-5.40	20.00	10.30	8.30
10	No MTB	No FAT	10	-10.94	-5.26	-2.73	-6.31
11	No MTB	No FAT	11	-4.19	-11.83	-6.29	-7.44
12	No MTB	FAT present	1	88.16	97.58	66.27	84.00
13	No MTB	FAT present	2	196.50	114.10	134.20	148.27
14	No MTB	FAT present	3	6.02	1.19	3.38	3.53
15	No MTB	FAT present	4	935.40	1011.00	951.20	965.87
16	No MTB	FAT present	5	606.00	592.70	608.40	602.37
17	No MTB	FAT present	6	1457.00	1349.00	1625.00	1477.00
18	No MTB	FAT present	7	139.70	399.50	91.69	210.30
19	No MTB	FAT present	8	196.70	270.80	160.70	209.40
20	No MTB	FAT present	9	135.20	221.50	268.00	208.23

21	No MTB	FAT present	10	-14.47	79.62	304.10	123.08
22	No MTB	FAT present	11	516.30	585.90	562.60	554.93
23	MTB present	No FAT	1	-0.05	-0.09	-0.08	-0.07
24	MTB present	No FAT	2	-9.41	-6.30	5.40	-3.44
25	MTB present	No FAT	3	-10.13	-16.48	-14.70	-13.77
26	MTB present	No FAT	4	8.75	134.90	203.70	115.78
27	MTB present	No FAT	5	612.30	898.20	854.20	788.23
28	MTB present	No FAT	6	2034.00	2743.00	2772.00	2516.33
29	MTB present	No FAT	7	978.50	1137.00	850.00	988.50
30	MTB present	No FAT	8	279.30	124.80	222.10	208.73
31	MTB present	No FAT	9	688.10	530.90	720.20	646.40
32	MTB present	No FAT	10	908.80	811.90	746.40	822.37
33	MTB present	No FAT	11	439.30	960.90	593.30	664.50
34	MTB present	FAT present	1	709.30	874.30	630.00	737.87
35	MTB present	FAT present	2	4541.00	4106.00	4223.00	4290.00
36	MTB present	FAT present	3	391.00	194.00	254.00	279.67
37	MTB present	FAT present	4	2913.00	3632.00	3417.00	3320.67
38	MTB present	FAT present	5	3801.00	3112.00	3681.00	3531.33
39	MTB present	FAT present	6	10150.00	9410.00	9243.00	9601.00
40	MTB present	FAT present	7	6736.00	6323.00	5117.00	6058.67
41	MTB present	FAT present	8	1454.00	2250.00	1092.00	1598.67
42	MTB present	FAT present	9	857.20	1339.00	1945.00	1380.40
43	MTB present	FAT present	10	.	739.90	4379.00	.
44	MTB present	FAT present	11	6637.00	6909.00	6453.00	6666.33

2.6   PROC IMPORT

SAS can use this procedure to read in data from files that have special structures. These can be plain text files in which the variables are separated by delimiters (see below for further details) or files containing data created by another software package.

2.6.1   DELIMITED FILES

A delimited file is the simplest type of file that can be accessed using PROC IMPORT. This is a file in which each row of data has its variables separated by a particular character, such as a comma, the TAB character, or some other character. When a comma is used, the data are often referred to as being stored as "comma-separated values." When a TAB character is used, they are often referred to as being stored as "tab-delimited values."

The TNF data shown in Table 2.1 are available in a plain text file "tnfdatadelim.dat", where the "#" character is used as a delimiter, separating the variables. Part of this file is shown below. Missing data have no values between two "#" delimiters.

```
mtb#fat#donor#rep1#rep2#rep3
0#0#1#-0.01#-0.01#-0.13
0#0#2#16.13#-9.62#-14.8
0#0#3##-0.3#-0.95
0#0#4#3.63#47.5#55.2
 .
 .
 .
1#1#8#1454#2250#1092
1#1#9#857.2#1339#1945
1#1#10##739.9#4379
1#1#11#6637#6909#6453
```

Here, the variable names are contained in the first line of the file, and using the GETNAMES = YES statement in the PROC

IMPORT will recognise them as such. If they were omitted or invalid
SAS names, then the names "var1", "var2", etc., would be used by
SAS.

Program 2.6.1 may be used to read this data file using PROC
IMPORT. The PROC PRINT procedure is used to show the data. With
the REPLACE option, any dataset that already exists with the name
defined by the OUTFILE statement will be overwritten.

PROGRAM 2.6.1

```
Comments omitted

PROC IMPORT DATAFILE =
  'a:\data\tnfdatadelim.dat'
  OUT = tnfdata DBMS = DLM REPLACE;
DELIMITER = '#';
GETNAMES = YES;
```
*The delimited file is imported, and the SAS dataset created is
given the name "tnfdata". DBMS = DLM tells SAS that it is a
delimited file, and the DELIMITER statement tells SAS what
character is used as the delimiter. The GETNAMES = YES
statement reads in the first line of the file as variable names.*

```
PROC PRINT DATA = tnfdata;
```
The PROC PRINT displays the "tnfdata" dataset.

```
RUN;
```

SAS Output for Program 2.6.1

obs	mtb	fat	donor	rep1	rep2	rep3
1	0	0	1	−0.01	−0.01	−0.13
2	0	0	2	16.13	−9.62	−14.8
3	0	0	3	.	−0.3	−0.95
4	0	0	4	3.63	47.5	55.2

5	0	0	5	-3.21	-5.64	-5.32
6	0	0	6	16.26	52.21	17.93
7	0	0	7	-12.74	-5.23	-4.06
8	0	0	8	-4.67	20.1	110
9	0	0	9	-5.4	20	10.3
10	0	0	10	-10.94	-5.26	-2.73
11	0	0	11	-4.19	-11.83	-6.29
12	0	1	1	88.16	97.58	66.27
13	0	1	2	196.5	114.1	134.2
14	0	1	3	6.02	1.19	3.38
15	0	1	4	935.4	1011	951.2
16	0	1	5	606	592.7	608.4
17	0	1	6	1457	1349	1625
18	0	1	7	139.7	399.5	91.69
19	0	1	8	196.7	270.8	160.7
20	0	1	9	135.2	221.5	268
21	0	1	10	-14.47	79.62	304.1
22	0	1	11	516.3	585.9	562.6
23	1	0	1	-0.05	-0.09	-0.08
24	1	0	2	-9.41	-6.3	5.4
25	1	0	3	-10.13	-16.48	-14.7
26	1	0	4	8.75	134.9	203.7
27	1	0	5	612.3	898.2	854.2
28	1	0	6	2034	2743	2772
29	1	0	7	978.5	1137	850
30	1	0	8	279.3	124.8	222.1
31	1	0	9	688.1	530.9	720.2
32	1	0	10	908.8	811.9	746.4
33	1	0	11	439.3	960.9	593.3
34	1	1	1	709.3	874.3	630
35	1	1	2	4541	4106	4223
36	1	1	3	391	194	254
37	1	1	4	2913	3632	3417
38	1	1	5	3801	3112	3681
39	1	1	6	10150	9410	9243

40	1	1	7	6736	6323	5117
41	1	1	8	1454	2250	1092
42	1	1	9	857.2	1339	1945
43	1	1	10	.	739.9	4379
44	1	1	11	6637	6909	645

2.6.2 SPREADSHEET FILES

PROC IMPORT can read data directly from most common versions of
Microsoft Excel and Lotus 1-2-3® spreadsheet packages. As an example
here, the TNF data shown in Table 2.1 are available in a Microsoft Excel
97 workbook "tnfdata.xls", which contains the data with the variable
names in the first row.

Program 2.6.2 may be used to read this data file using PROC
IMPORT. The PROC PRINT procedure is used to show the data.

PROGRAM 2.6.2

```
Comments omitted

PROC IMPORT DATAFILE =
   'a:\data\tnfdata.xls' OUT = tnfdata
  DBMS = EXCEL97 REPLACE;
SHEET = TNFdata;
GETNAMES = YES;
```
*The Microsoft Excel file is imported, and the SAS dataset cre-
ated is given the name "tnfdata". DBMS = EXCEL97 tells SAS
that it is a Microsoft Excel 97 file, and the SHEET statement
tells SAS the worksheet of the workbook to import. REPLACE
means that any dataset already called "tnfdata" will be over-
written. The "GETNAMES = YES" statement reads in the first
line of the file as variable names.*

```
PROC PRINT DATA = tnfdata;
```
The PROC PRINT displays the "tnfdata" dataset.

```
RUN;
```

SAS Output for Program 2.6.2

This is the same as the output for Program 2.6.1.

Additional Notes for Program 2.6.2

DBMS = EXCEL97 — Here, the version of Microsoft Excel used to create the file is specified. However, SAS is able to discern the different versions of Microsoft Excel, and the statement DBMS = EXCEL would have been sufficient. This feature could be useful if the version of Microsoft Excel used to create the file is unknown. In fact, the DBMS statement could have been omitted, because SAS would recognise the ".xls" file extension and assume that a Microsoft Excel file was being imported.

SHEET = TNFdata; — The workbook being imported ("tnfdata.xls") contains only this worksheet called "TNFdata", so this statement could have been omitted.

2.6.3 DATABASE FILES

PROC IMPORT can read data directly from most common versions of Microsoft Access® and dBase® database packages. As an example, here the TNF data shown in Table 2.1 are available in a Microsoft Access 97 database "tnfdata.mdb", which has an additional "ID" variable as a primary key for the database.

Program 2.6.3 may be used to read this data file using PROC IMPORT. The PROC PRINT procedure is used to show the data.

PROGRAM 2.6.3

```
Comments omitted

PROC IMPORT TABLE = 'TNFdata' OUT = tnfdata
  DBMS = ACCESS97 REPLACE;
DATABASE = 'a:\data\tnfdata.mdb';
```

The Microsoft Access file is imported, and the SAS dataset created is given the name "tnfdata". DBMS = ACCESS97 tells SAS that it is a Microsoft Access 97 file, and the TABLE option tells SAS which table of the database to import. REPLACE means that any dataset already called "tnfdata" will be over-written. The DATABASE statement tells SAS the name and location of the file.

```
PROC PRINT DATA = tnfdata;
```
The PROC PRINT displays the "tnfdata" dataset.

```
RUN;
```

SAS Output for Program 2.6.3

Obs	ID	mtb	fat	donor	rep1	rep2	rep3
1	1	0	0	1	−0.01	−0.01	−0.13
2	2	0	0	2	16.13	−9.62	−14.80
3	3	0	0	3	.	−0.30	−0.95
4	4	0	0	4	3.63	47.50	55.20
5	5	0	0	5	−3.21	−5.64	−5.32
6	6	0	0	6	16.26	52.21	17.93
7	7	0	0	7	−12.74	−5.23	−4.06
8	8	0	0	8	−4.67	20.10	110.00
9	9	0	0	9	−5.40	20.00	10.30
10	10	0	0	10	−10.94	−5.26	−2.73
11	11	0	0	11	−4.19	−11.83	−6.29
12	12	0	1	1	88.16	97.58	66.27
13	13	0	1	2	196.50	114.10	134.20
14	14	0	1	3	6.02	1.19	3.38
15	15	0	1	4	935.40	1011.00	951.20
16	16	0	1	5	606.00	592.70	608.40
17	17	0	1	6	1457.00	1349.00	1625.00
18	18	0	1	7	139.70	399.50	91.69

19	19	0	1	8	196.70	270.80	160.70
20	20	0	1	9	135.20	221.50	268.00
21	21	0	1	10	−14.47	79.62	304.10
22	22	0	1	11	516.30	585.90	562.60
23	23	1	0	1	−0.05	−0.09	−0.08
24	24	1	0	2	−9.41	−6.30	5.40
25	25	1	0	3	−10.13	−16.48	−14.70
26	26	1	0	4	8.75	134.90	203.70
27	27	1	0	5	612.30	898.20	854.20
28	28	1	0	6	2034.00	2743.00	2772.00
29	29	1	0	7	978.50	1137.00	850.00
30	30	1	0	8	279.30	124.80	222.10
31	31	1	0	9	688.10	530.90	720.20
32	32	1	0	10	908.80	811.90	746.40
33	33	1	0	11	439.30	960.90	593.30
34	34	1	1	1	709.30	874.30	630.00
35	35	1	1	2	4541.00	4106.00	4223.00
36	36	1	1	3	391.00	194.00	254.00
37	37	1	1	4	2913.00	3632.00	3417.00
38	38	1	1	5	3801.00	3112.00	3681.00
39	39	1	1	6	10150.00	9410.00	9243.00
40	40	1	1	7	6736.00	6323.00	5117.00
41	41	1	1	8	1454.00	2250.00	1092.00
42	42	1	1	9	857.20	1339.00	1945.00
43	43	1	1	10	.	739.90	4379.00
44	44	1	1	11	6637.00	6909.00	6453.00

2.7 READING DATES AND TIMES

Variables that are dates and times can have special formats attached to them so that SAS knows to treat them accordingly. One way of assigning these formats to variables is to define them in the INPUT statement of the data step.

The TNF data in Table 2.1 show the addition of date and time variables to the file "tnfdatadatestimes.dat", which may relate the date and time the data were entered into a computer system. Part of this file is shown below, illustrating that, for example, the first line of data is associated with the date 07/05/03 and the time 14:07:

```
001   -0.01  -0.01  -0.13 07/05/03 14:07
002   16.13  -9.62 -14.8  06/05/03 14:51
003                 -0.3  -0.95 08/05/03 11:51
004    3.63   47.5  55.2  10/05/03 11:27
.
.
.
118    1454   2250  1092  06/05/03 12:57
119   857.2   1339  1945  06/05/03 18:31
1110          739.9 4379  06/05/03 11:28
1111  6637    6909  6453  07/05/03 17:18
```

Program 2.7.3 reads in this data, assigning special formats to the date and time variables. The PROC PRINT procedure is used to show the data.

PROGRAM 2.7.1

```
Comments omitted

DATA tnfdata;
INFILE 'a:\data\tnfdatadatestimes.dat'
  TRUNCOVER;
INPUT mtb 1 fat 2 donor 3-4 rep1 5-10
  rep2 11-16 rep3 17-21 datein:DDMMYY8.
  timein:TIME5.;
FORMAT datein DDMMYY8. timein TIME5.;
```
The format DDMMYY8 is associated with the "datein" variable, and the format TIME5 is associated with the "timein" variable.

```
PROC PRINT DATA = tnfdata;
VAR mtb fat donor datein timein;
The PROC PRINT displays the specified variables from the
"tnfdata" dataset.

RUN;
```

SAS Output for Program 2.7.1

Obs	mtb	fat	donor	datein	timein
1	0	0	1	07/05/03	14:07
2	0	0	2	06/05/03	14:51
3	0	0	3	08/05/03	11:51
4	0	0	4	10/05/03	11:27
5	0	0	5	10/05/03	18:49
6	0	0	6	10/05/03	12:32
7	0	0	7	06/05/03	17:44
8	0	0	8	08/05/03	10:20
9	0	0	9	10/05/03	15:23
10	0	0	10	06/05/03	16:55
11	0	0	11	07/05/03	18:17
12	0	1	1	06/05/03	18:46
13	0	1	2	06/05/03	13:44
14	0	1	3	06/05/03	16:29
15	0	1	4	07/05/03	10:58
16	0	1	5	06/05/03	14:44
17	0	1	6	07/05/03	14:03
18	0	1	7	07/05/03	18:52
19	0	1	8	08/05/03	17:55
20	0	1	9	07/05/03	10:26
21	0	1	10	07/05/03	18:39
22	0	1	11	07/05/03	13:52
23	1	0	1	10/05/03	18:51
24	1	0	2	08/05/03	9:58
25	1	0	3	08/05/03	19:02

26	1	0	4	07/05/03	10:03
27	1	0	5	10/05/03	13:18
28	1	0	6	10/05/03	19:13
29	1	0	7	10/05/03	15:24
30	1	0	8	07/05/03	9:48
31	1	0	9	10/05/03	16:52
32	1	0	10	06/05/03	12:53
33	1	0	11	09/05/03	12:40
34	1	1	1	10/05/03	15:50
35	1	1	2	10/05/03	16:30
36	1	1	3	08/05/03	13:15
37	1	1	4	07/05/03	13:17
38	1	1	5	09/05/03	16:19
39	1	1	6	07/05/03	15:29
40	1	1	7	09/05/03	13:40
41	1	1	8	06/05/03	12:57
42	1	1	9	06/05/03	18:31
43	1	1	10	06/05/03	11:28
44	1	1	11	07/05/03	17:18

Additional Notes for Program 2.7.1

```
INPUT mtb 1 fat 2 donor 3-4 rep1 5-10 rep2
11-16   rep3   17-21   datein:DDMMYY8.
timein:TIME5.;
```
— The format is attached to the "datein"
variable by having a colon following the variable name, then
the code for the format (DDMMYY8), and then a full stop. The
DDMMYY8 relates to a date that is eight characters wide with
days first (DD), then months (MM), and then years (YY). Similarly, the TIME5 format associated with the "timein" variable
relates to a time that is five characters wide, with hours first,
then a colon, then the minutes.

2.8   TASKS

These tasks refer to the "Nurses' Glove Use Data" in Appendix A.

Task 2.1

Construct a SAS program that does the following:

1. Reads in the "gloves.dat" data
2. Applies appropriate formats for the period variable and variable labels for all variables
3. Creates a new variable for the proportion of observations in which gloves are used (number of times gloves used/total number of times observed)
4. Uses PROC PRINT to list the dataset

Task 2.2

1. Using a text editor, type in part of the Nurses' Glove Use Dataset, using the TAB character and spaces arbitrarily to put the variables in neat columns. This will give a good idea of how a rectangular data file can be produced with relative ease but little care.
2. Copy and amend the SAS program from Task 2.1 to try and read in the data file created in (1), above. Look at the log file in conjunction with the data file to try and ascertain what is happening.

Task 2.3

Run Program 2.1.1 without the TRUNCOVER option to see the effect it has.

TASK 2.4

Repeat Task 2.1 but use PROC IMPORT to access the data in
the file "glovesdelim.dat", which is delimited by semicolons,
instead of reading in "gloves.dat".

3

Saving and Output of SAS Datasets

Rather than having to read in raw data or use PROC IMPORT and then apply labels and formats and create new variables each time an analysis of a dataset is required, SAS enables the user to save a dataset. This creation of a permanent dataset will preserve the labels, formats, etc. Future programs can then access this saved dataset directly for further analysis. The creation and use of these datasets are discussed in Section 3.1 and Section 3.2.

However, a SAS permanent dataset is not something that can be examined and understood by the user without using SAS. For this to happen, the dataset must be output in some manner. Here, three output methods are discussed: PROC REPORT, the PUT statement, and PROC EXPORT.

3.1 PERMANENT DATASETS

In Chapter 2, the TNF dataset was read into SAS as dataset "tnfdata", and average TNF ("tnfaver") was calculated. This "tnfdata" dataset is then available with "tnfaver" throughout that SAS session. However, once that SAS session terminates (e.g., by exiting the software), then this "temporary" dataset is discarded, along with the "tnfaver" variable. In a simple situation as presented here, this is not a major problem — if further analysis of the TNF data is required, it can all be read into SAS again, and "tnfaver" can be recalculated. It may be only slightly inconvenient, but it is sometimes the case that complex manipulations are carried out on datasets, and it would be wasteful to have to repeat these each time the manipulated dataset was required for analysis.

SAS overcomes this problem by being able to save "permanent" datasets. To do this, SAS thinks of the datasets as being contained in different libraries. This can be likened to storing computer files in different folders on a disk. SAS thus needs to know two things: where to store the permanent dataset (disk drive and folder, if necessary) and what name to call the dataset.

In Program 3.1.1, the first LIBNAME statement defines "abcxyz" as being the path "a:\data". This needs to exist in the program, because SAS needs something to act as a placeholder for the "a:\data" when it comes to the DATA step. Anything can be used instead of "abcxyz" (subject to certain constraints; SAS will produce an error message if constraints are violated).

The second LIBNAME statement in Program 3.1.1 defines "library" as being the path "a:\formats". SAS saves the formats separately from the actual data and, using the specific word "library" as the placeholder, enables SAS to locate them when they are applied to variables using the FORMAT statement.

PROGRAM 3.1.1

```
Comments omitted

LIBNAME abcxyz 'a:\data';
LIBNAME library 'a:\formats';

PROC FORMAT LIBRARY = library.formats;
VALUE mtblab 0 = 'No MTB'
  1 = 'MTB present';
VALUE fatlab 0 = 'No FAT'
  1 = 'FAT present';

DATA abcxyz.tnfdata;
INFILE 'a:\data\tnfdata.dat' TRUNCOVER;
INPUT mtb 1 fat 2 donor 3-4 rep1 5-10
  rep2 11-16 rep3 17-22;
LABEL rep1 = '1st replicate';
LABEL rep2 = '2nd replicate';
LABEL rep3 = '3rd replicate';
LABEL mtb =  'MTB presence';
```

```
LABEL  fat = 'FAT presence';
tnfaver = (rep1+rep2+rep3)/3;
LABEL tnfaver = 'average TNF';
FORMAT mtb mtblab.; FORMAT fat fatlab.;
RUN;
```

SAS Output for Program 3.1.1

There is no output from Program 3.1.1, apart from the permanent dataset.

In Program 3.1.1, the data, including the new "tnfaver", along with the labels are saved as file "a:\data\tnfdata.sas7bdat". Formats are saved as file "a:\formats\formats.sas7bcat". The dataset and formats can be accessed at any point in the future, as required. Note that the directory "a:\formats" must already exist.

To help with the identification of permanent datasets, there is a procedure PROC CONTENTS, shown in Program 3.1.2.

PROGRAM 3.1.2

```
Comments omitted

LIBNAME  abcxyz 'a:\data';
LIBNAME  library 'a:\formats';
PROC CONTENTS DATA = abcxyz.tnfdata;
RUN;
```

The first LIBNAME statement again defines "abcxyz" as being the path "a:\data". The second use of LIBNAME tells SAS where to look for the formats that were saved. The PROC CONTENTS can then directly access the permanent dataset "tnfdata".

SAS Output for Program 3.1.2

The CONTENTS Procedure

Data Set Name:	ABCXYZ.TNFDATA	Observations:	44
Member Type:	DATA	Variables:	7
Engine:	V8	Indexes:	0
Created:	[date of creation]	Observation Length:	56
Last Modified:	[date of last modification]	Deleted Observations:	0
Protection:		Compressed:	NO
Data Set Type:		Sorted:	NO
Label:			

- - - Engine/Host Dependent Information - - -

Data Set Page Size:	8192
Number of Data Set Pages:	1
First Data Page:	1
Max Obs per Page:	145
Obs in First Data Page:	44
Number of Data Set Repairs:	0
File Name:	a:\data\tnfdata.sas7bdat
Release Created:	8.0202M0
Host Created:	WIN_95

```
- - - Alphabetic List of Variables and Attributes - - -
```

#	Variable	Type	Len	Pos	Format	Label
3	donor	Num	8	16		
2	fat	Num	8	8	FATLAB.	FAT presence
1	mtb	Num	8	0	MTBLAB.	MTB presence
4	rep1	Num	8	24		1st replicate
5	rep2	Num	8	32		2nd replicate
6	rep3	Num	8	40		3rd replicate
7	tnfaver	Num	8	48		average TNF

3.2   DATA MANIPULATION, PERMANENT DATASETS, AND THE SET STATEMENT

Once a permanent dataset is saved, it is possible that future programs will want to carry out further manipulations on it. In order to do so, it is good practice to create a temporary dataset based on the permanent dataset and manipulate the temporary dataset, as shown in Program 3.2.1, thereby avoiding any potential difficulties with SAS trying to overwrite datasets stored on disk.

PROGRAM 3.2.1

Comments omitted

```
LIBNAME abcxyz 'a:\data';
LIBNAME library 'a:\formats';

DATA tnf;
SET abcxyz.tnfdata;
.
.
```
More lines of data manipulation and analysis procedures
```
.
.
```
*This creates a temporary dataset, "tnf", and the SET state-
ment defines it to be the same as the permanent dataset before
further manipulations are undertaken.*
```
RUN;
```

3.3 PROC REPORT

This procedure (shown in Program 3.3.1) offers more control over the
listing of data than PROC PRINT.

PROGRAM 3.3.1

Comments omitted

```
LIBNAME abcxyz 'a:\data';
LIBNAME library 'a:\formats';

PROC REPORT DATA = abcxyz.tnfdata;
COLUMN mtb fat tnfaver;
TITLE 'TNF Data';
DEFINE mtb/ORDER WIDTH = 12 'MTB presence'
   F = 1.0;
```

```
DEFINE  fat/ORDER  WIDTH = 12
   'FAT  presence 'F = 1.0;
DEFINE tnfaver/WIDTH = 12 'average TNF'
   F = 8.3;
```

The COLUMN statement specifies the variables to be reported. The columns are then defined. There are many possible options (shown after the "/": ORDER suppresses continual output of repeated values; WIDTH specifies how wide the column should be; a heading for the column is given inside inverted commas; a format is defined (e.g., F = 8.3 means that the number output is to be eight characters wide with three decimal places).

```
RUN;
```

The output from PROC REPORT opens in a special "Report" window (unless options are used that stop this from happening). The menus and toolbars for this window allow for additional editing of the report. This more advanced use of SAS is beyond the scope of this book.

3.4 PUT

The PUT statement can be used to output data to a text file. It must be used within a data step. Shown in Program 3.4.1 is an example.

Program 3.4.1

```
Comments omitted

LIBNAME  abcxyz  'a:\data';
LIBNAME  library  'a:\formats';

DATA tnf;
SET abcxyz.tnfdata;
```

```
FILE 'a:\tnf.dat';
PUT mtb 1  fat  3  tnfaver  5-11.2;
```
*The PUT statement outputs the variable "mtb" to Column 1
of the file "a:\tnf.dat", the variable "fat" to Column 3, and
the variable "tnfaver" to Columns 5 to 11, with two decimal
places (caused by the ".2").*

```
RUN;
```

In this case, the creation of the temporary dataset "tnf" is wasteful — it is not used after the PUT statement is used. In these
circumstances, the line DATA _NULL_; could be used instead
of the existing line DATA TNF;. This would not create a temporary dataset but would still allow the use of the PUT statement.

File "a:\tnf.dat" from Program 3.4.1

```
0 0 -0.05
0 0 -2.76
0 0 .
0 0 35.44
. . .
. . .
. . .
1 1 1598.67
1 1 1380.40
1 1 .
1 1 6666.33
```

3.5   PROC EXPORT

Just as PROC IMPORT can import data from files that have a special
structure, PROC EXPORT can take a SAS dataset and from it, create a
file with a special structure. This structure may simply be a delimited

text file or may be a file that can be read by another software package. Several types of files can be exported by SAS, an up-to-date list of which can be found by consulting the SAS manuals or help facilities.

3.5.1 DELIMITED FILES

Program 3.5.1 takes the TNF dataset and exports it to a text file with the delimiter "%". With the REPLACE option, any previous file with the name defined by the OUTFILE statement will be overwritten.

PROGRAM 3.5.1

Comments omitted

```
LIBNAME abcxyz 'a:\data';
LIBNAME library 'a:\formats';
PROC EXPORT DATA = abcxyz.tnfdata
   DBMS = DLM
OUTFILE = 'a:\tnfdatadelim2.dat' REPLACE;
DELIMITER = '%';
```
The delimited file "a:\tnfdatadelim2.dat" is exported, based on the dataset defined by the DATA = option. DBMS = DLM tells SAS that it is a delimited file, and the DELIMITER statement tells SAS what character is used as the delimiter.

```
RUN;
```

File "a:\tnfdatadelim2.dat" from Program 3.5.1

```
mtb%fat%donor%rep1%rep2%rep3%tnfaver
No MTB%No FAT%1%-0.01%-0.01%-0.13%-0.05
No MTB%No FAT%2%16.13%-9.62%-14.8%
   2.763333333
No MTB%No FAT%3%%-0.3%-0.95%
No MTB%No FAT%4%3.63%47.5%55.2%
   35.443333333
```

```
.%.%.% .%.% .%.
.%.%.% .%.% .%.
.%.%.% .%.% .%.
MTB present%FAT  present%8%1454%2250%1092%
  1598.6666667
MTB present%FAT  present%9%857.2%1339%
  1945%1380.4
MTB present%FAT  present%10%%739.9%4379%
MTB present%FAT  present%11%6637%6909%
  6453%6666.3333333
```

3.5.2 SPREADSHEET FILES

PROC EXPORT can output a SAS dataset to a file that can be read by most common versions of Microsoft Excel and Lotus 1-2-3 spreadsheet packages. To do this, the DBMS option in Program 3.5.1 would have to be defined appropriately. Care must be taken when exporting to a Microsoft Excel file. The option DBMS = EXCEL will work, but the resulting file will be of a format that cannot be read by early versions of Microsoft Excel. More specific choices, such as DBMS = EXCEL97 (for Microsoft Excel 97), can be used instead.

3.5.3 DATABASE FILES

In a similar manner to spreadsheet files, PROC EXPORT can output an SAS dataset to a file that can be read by most common versions of Microsoft Access and dBase database packages. The DBMS option in Program 3.5.1 would have to be defined appropriately. Again, care must be taken when exporting to a Microsoft Access file. The option DBMS = ACCESS will work, but the resulting file will be of a format that cannot be read by early versions of Microsoft Access. More specific choices, such as DBMS = ACCESS97 (for Microsoft Access 97), can be used instead.

3.6 TASKS

These tasks refer to the Nurses' Glove Use Data in Appendix A.

TASK 3.1

Copy and amend the SAS program from Task 2.1 to save the data as a permanent dataset. Also, save the formats used.

TASK 3.2

Create a SAS program using PROC CONTENTS that examines the dataset saved in Task 3.1.

TASK 3.3

Refer to the saved versions of the Nurses' Glove Use Data and formats (created in Task 3.1), and use PROC REPORT to produce a neat listing of the data. Note that you will need to close the report that is produced before doing further analyses.

TASK 3.4

Output the Nurses' Glove Use Data (saved as a permanent dataset in Task 3.1) to a text file using the PUT statement.

TASK 3.5

Use PROC EXPORT to output the Nurses' Glove Use Data from a SAS dataset to a delimited text file, using a delimiter of your choice.

4 Manipulating Datasets

It is often the case that data collected on different occasions or in different locations are contained in separate datasets. Before any meaningful analysis can take place, these datasets need to be merged and concatenated (collated together).

4.1 MERGING DATASETS

The TNF data in Table 2.1 come from three replicates of four different experiments: (1) MTB and FAT not present; (2) MTB not present, FAT present; (3) MTB present, FAT not present; and (4) MTB and FAT present. Taking the first of these, the data collected for the three replications may have been recorded in three files, as illustrated in Table 4.1. Note that the reading for the first replicate of Donor 3 is missing.

The first step of putting the TNF dataset together is to merge these three datasets. We can use the donor number to link the cases. Program 4.1.1 performs this in SAS and saves the result as a permanent dataset for use in Section 4.2. Program 4.1.2 through Program 4.1.4 (available electronically via the web site mentioned in the preface but not printed here) perform this task for the other three possible combinations of MTB and FAT being present or not present.

Program 4.1.1

```
Comments omitted
LIBNAME abcxyz 'a:\data';
DATA tnfdata11;
INFILE 'a:\data\tnfdata11.dat'
    TRUNCOVER;
INPUT donor 1-2 rep1 3-8;
```

TABLE 4.1
TNF Data before Merging

First Replicate		Second Replicate		Third Replicate	
Donor	TNF	Donor	TNF	Donor	TNF
1	-0.01	1	-0.01	1	-0.13
2	16.13	2	-9.62	2	-14.88
4	3.63	3	-0.3	3	-0.95
5	-3.21	4	47.5	4	55.2
6	16.26	5	-5.64	5	-5.32
7	-12.74	6	52.21	6	17.93
8	-4.67	7	-5.23	7	-4.06
9	-5.4	8	20.1	8	110
10	-10.94	9	20	9	10.3
11	-4.19	10	-5.26	10	-2.73
		11	-11.83	11	-6.29

```
PROC SORT; BY donor;
```
The PROC SORT sets up the variable "donor" as a key for the merging.
```
DATA tnfdata12;
INFILE 'a:\data\tnfdata12.dat'
  TRUNCOVER;
INPUT donor 1-2 rep2 3-8;
PROC SORT; BY donor;
```
The PROC SORT sets up the variable "donor" as a key for the merging.
```
DATA tnfdata13;
INFILE 'a:\data\tnfdata13.dat'
  TRUNCOVER;
INPUT donor 1-2 rep3 3-8;
```

```
PROC SORT; BY donor;
```
The PROC SORT sets up the variable "donor" as a key for the merging.

```
DATA abcxyz.tnfdata1;
MERGE tnfdata11 tnfdata12 tnfdata13;
BY donor;
```
All three datasets are merged together using the variable "donor" as key.

```
PROC PRINT DATA=abcxyz.tnfdata1;
RUN;
```

SAS Output for Program 4.1.1

Obs	donor	rep1	rep2	rep3
1	1	−0.01	−0.01	−0.13
2	2	16.13	−9.62	−14.88
3	3	.	−0.30	−0.95
4	4	3.63	47.50	55.20
5	5	−3.21	−5.64	−5.32
6	6	16.26	52.21	17.93
7	7	−12.74	−5.23	−4.06
8	8	−4.67	20.10	110.00
9	9	−5.40	20.00	10.30
10	10	−10.94	−5.26	−2.73
11	11	−4.19	−11.83	−6.29

Additional Notes for Program 4.1.1

PROC SORT; BY donor; — In Program 4.1, the donor
   number is the variable that occurs in each of the datasets,
   "tnfdat11", "tnfdat12", and "tnfdat13", that SAS needs to

look at in order to work out which case of "tnfdat11"
matches which cases of "tnfdat12" and "tnfdat13". Because
of this, each of these datasets must be sorted by the donor
number (with PROC SORT) before being merged in the final
data step with the statement BY DONOR.

MERGE tnfdata11 (IN=t1) tnfdata12 (IN=t2)
tnfdata13 (IN=t3); — This is an alternative to the
simple MERGE statement used. It defines three additional
logical variables. Variable "t1" is TRUE for cases that exist
in dataset "tnfdat11" (i.e., it will be TRUE for Donors 1, 2,
and 4 through 11, but FALSE for Donor 3, which has no
record in this dataset). Variables "t2" and "t3" are similarly
defined.

If the statement "IF t1 AND t2 AND t3;" were included in
the data step, then the resulting merged dataset would
only contain cases that were TRUE for all of "t1", "t2",
and "t3". In this case, this would be Donors 1, 2, and 4
through 11 (Donor 3 being excluded as a result of not
having a record in dataset "tnfdat11").

If the statement "IF t1;" were included in the data step, then
the resulting merged dataset would only contain cases
that were TRUE for "t1". In this case, this would again
be Donors 1, 2, and 4 through 11 (Donor 3 being excluded as a result of not having a record in dataset
"tnfdat11").

If the statement "IF t2;" were included in the data step, then
the resulting merged dataset would only contain cases
that were TRUE for "t2". In this case, this would be Donors 1 through 11 (Donor 3 now being included, because a record for it exists in dataset "tnfdat12").

Various combinations of these logical variables and the
Boolean operators AND, OR, and NOT give great flexibility to the merging process.

4.2 CONCATENATING DATASETS

For each of the four possible combinations of MTB and FAT being present or not present, procedures such as those in Section 4.1 have been carried out to merge the datasets. The TNF data is then in the form shown in Table 4.2.

Each of the datasets in Table 4.2 was saved as a permanent SAS dataset. In Program 4.2.1, they are read in separately and then joined.

PROGRAM 4.2.1

Comments omitted

```
LIBNAME abcxyz 'a:\data';
DATA tnf1; SET abcxyz.tnfdata1;
mtb=0; fat=0;
```
A new dataset, "tnf1" is created, initially based on the permanent dataset "abcxyz.tnfdata1". The appropriate values for "mtb" and "fat" are added to this new dataset.

```
DATA tnf2; SET abcxyz.tnfdata2;
mtb=1; fat=0;
```
A new dataset "tnf2" is created, initially based on the permanent dataset "abcxyz.tnfdata2". The appropriate values for "mtb" and "fat" are added to this new dataset.

```
DATA tnf3; SET abcxyz.tnfdata3;
mtb=0; fat=0;
```
A new dataset "tnf3" is created, initially based on the permanent dataset "abcxyz.tnfdata3". The appropriate values for "mtb" and "fat" are added to this new dataset.

```
DATA tnf4; SET abcxyz.tnfdata4;
mtb=0; fat=1;
```
A new dataset "tnf4" is created, initially based on the permanent dataset "abcxyz.tnfdata4". The appropriate values for "mtb" and "fat" are added to this new dataset.

TABLE 4.2
TNF Data before Concatenating

TNF Dataset 1 (MTB Not Present, FAT Not Present)				TNF Dataset 2 (MTB Present, FAT Not Present)			
Donor	TNF, Three Replicates			Donor	TNF, Three Replicates		
1	-0.01	-0.01	-0.13	1	-0.05	-0.09	-0.08
2	16.13	-9.62	-14.88	2	-9.41	-6.3	5.4
3	Missing	-0.3	-0.95	3	-10.13	-16.48	-14.79
4	3.63	47.5	55.2	4	8.75	134.9	203.7
5	-3.21	-5.64	-5.32	5	612.3	898.2	854.2
6	16.26	52.21	17.93	6	2034	2743	2772
7	-12.74	-5.23	-4.06	7	978.5	1137	850
8	-4.67	20.1	110	8	279.3	124.8	222.1
9	-5.4	20	10.3	9	688.1	530.9	720.2
10	-10.94	-5.26	-2.73	10	908.8	811.9	746.4
11	-4.19	-11.83	-6.29	11	439.3	960.9	593.3

TNF Dataset 3 (MTB Not Present, FAT Present)				TNF Dataset 4 (MTB Present, FAT Present)			
Donor	TNF, Three Replicates			Donor	TNF, Three Replicates		
1	88.16	97.58	66.27	1	709.3	874.3	630
2	196.5	114.1	134.2	2	4541	4106	4223
3	6.02	1.19	3.38	3	391	194	254
4	935.4	1011	951.2	4	2913	3632	3417
5	606	592.7	608.4	5	3801	3112	3681
6	1457	1349	1625	6	10150	9410	9243
7	139.7	399.5	91.69	7	6736	6323	5117
8	196.7	270.8	160.7	8	1454	2250	1092
9	135.2	221.5	268	9	857.2	1339	1945
10	-14.47	79.62	304.1	10	Missing	739.9	4379
11	516.3	585.9	562.6	11	6637	6909	6453

```
DATA abcxyz.tnfdata;
SET tnf1 tnf2 tnf3 tnf4;
```
As all four datasets, "tnf1", "tnf2", "tnf3", and "tnf4", have the same variables, they can be concatenated by being listed after the SET statement.

```
PROC PRINT DATA=abcxyz.tnfdata;

RUN;
```

SAS Output for Program 4.2.1

Obs	donor	rep1	rep2	rep3	mtb	fat
1	1	-0.01	-0.01	-0.13	0	0
2	2	16.13	-9.62	-14.88	0	0
3	3	.	-0.30	-0.95	0	0
4	4	3.63	47.50	55.20	0	0
5	5	-3.21	-5.64	-5.32	0	0
6	6	16.26	52.21	17.93	0	0
7	7	-12.74	-5.23	-4.06	0	0
8	8	-4.67	20.10	110.00	0	0
9	9	-5.40	20.00	10.30	0	0
10	10	-10.94	-5.26	-2.73	0	0
11	11	-4.19	-11.83	-6.29	0	0
12	1	-0.05	-0.09	-0.08	1	0
13	2	-9.41	-6.30	5.40	1	0
14	3	-10.13	-16.48	-14.79	1	0
15	4	8.75	134.90	203.70	1	0
16	5	612.30	898.20	854.20	1	0
17	6	2034.00	2743.00	2772.00	1	0
18	7	978.50	1137.00	850.00	1	0
19	8	279.30	124.80	222.10	1	0
20	9	688.10	530.90	720.20	1	0

21	10	908.80	811.90	746.40	1	0
22	11	439.30	960.90	593.30	1	0
23	1	88.16	97.58	66.27	0	1
24	2	196.50	114.10	134.20	0	1
25	3	6.02	1.19	3.38	0	1
26	4	935.40	1011.00	951.20	0	1
27	5	606.00	592.70	608.40	0	1
28	6	1457.00	1349.00	1625.00	0	1
29	7	139.70	399.50	91.69	0	1
30	8	196.70	270.80	160.70	0	1
31	9	135.20	221.50	268.00	0	1
32	10	-14.47	79.62	304.10	0	1
33	11	516.30	585.90	562.60	0	1
34	1	709.30	874.30	630.00	1	1
35	2	4541.00	4106.00	4223.00	1	1
36	3	391.00	194.00	254.00	1	1
37	4	2913.00	3632.00	3417.00	1	1
38	5	3801.00	3112.00	3681.00	1	1
39	6	10150.00	9410.00	9243.00	1	1
40	7	6736.00	6323.00	5117.00	1	1
41	8	1454.00	2250.00	1092.00	1	1
42	9	857.20	1339.00	1945.00	1	1
43	10	.	739.90	4379.00	1	1
44	11	6637.00	6909.00	6453.00	1	1

Additional Notes for Program 4.2.1

DATA tnf1; SET abcxyz.tnfdata1; mtb=0;
fat=0; — This data step creates a dataset called "tnf1" in the WORK library (the default library, as no LIBNAME was used), which is based on the permanent dataset called "abcxyz.tnfdata1" (that is, the dataset called "tnfdata1", stored in the library defined by "abcxyz"). The two variables

"mtb" and "fat" are then created as equal to zero for the whole of this "tnf1" dataset. The code DATA abcxyz.tnfdata1; mtb=0; fat=0; could have been used instead, but this would then change the permanent dataset by the addition of the two new variables. To avoid this unnecessary (and possibly unwanted) alteration from occurring, the "tnf1" dataset was created, initially being made the same as the permanent dataset but then adding the new variables.

4.3   MERGING AND CONCATENATING IN ONE PROGRAM

The merging done in Program 4.1.1 to Program 4.1.4 and the concatenating done in Program 4.2.1 could have been done in one program without the need to save the four permanent datasets "abcxyz.tnfdata1", "abcxyz.tnfdata2", "abcxyz.tnfdata3", and "abcxyz.tnfdata4". This is shown in Program 4.3.1 that also creates and applies formats, gives variable labels, and creates a new variable that is the average of the three replicates.

PROGRAM **4.3.1**

```
Comments omitted

LIBNAME abcxyz 'a:\data';
LIBNAME library 'a:\formats';

PROC FORMAT LIBRARY=library.formats;
VALUE mtblab 0='No MTB' 1='MTB present';
VALUE fatlab 0='No FAT' 1='FAT present';
```

```
DATA tnfdata11;
INFILE 'a:\data\tnfdata11.dat'
  TRUNCOVER;
INPUT donor 1-2 rep1 3-8;
PROC SORT; BY donor;

DATA tnfdata12;
INFILE 'a:\data\tnfdata12.dat'
  TRUNCOVER;
INPUT donor 1-2 rep2 3-8;
PROC SORT; BY donor;

DATA tnfdata13;
INFILE 'a:\data\tnfdata13.dat'
  TRUNCOVER;
INPUT donor 1-2 rep3 3-8;
PROC SORT; BY donor;

DATA tnfdata1;
MERGE tnfdata11 tnfdata12 tnfdata13;
BY donor;
mtb=0; fat=0;
```

The merging of the data for the three replicates and the creation of the "mtb" and "fat" variables can all be done in the same data step.

```
DATA tnfdata21;
INFILE 'a:\data\tnfdata21.dat'
  TRUNCOVER;
INPUT donor 1-2 rep1 3-8;
PROC SORT; BY donor;

DATA tnfdata22;
INFILE 'a:\data\tnfdata22.dat'
  TRUNCOVER;
INPUT donor 1-2 rep2 3-8;
PROC SORT; BY donor;
```

```
DATA tnfdata23;
INFILE 'a:\data\tnfdata23.dat'
   TRUNCOVER;
INPUT donor 1-2 rep3 3-8;
PROC SORT; BY donor;
DATA tnfdata2;
MERGE tnfdata21 tnfdata22 tnfdata23;
BY donor;
mtb=1; fat=0;
DATA tnfdata31;
INFILE 'a:\data\tnfdata31.dat'
   TRUNCOVER;
INPUT donor 1-2 rep1 3-8;
PROC SORT; BY donor;
DATA tnfdata32;
INFILE 'a:\data\tnfdata32.dat'
   TRUNCOVER;
INPUT donor 1-2 rep2 3-8;
PROC SORT; BY donor;
DATA tnfdata33;
INFILE 'a:\data\tnfdata33.dat'
   TRUNCOVER;
INPUT donor 1-2 rep3 3-8;
PROC SORT; BY donor;
DATA tnfdata3;
MERGE tnfdata31 tnfdata32 tnfdata33;
BY donor;
mtb=0; fat=1;
DATA tnfdata41;
INFILE 'a:\data\tnfdata41.dat'
   TRUNCOVER;
```

```
INPUT donor 1-2 rep1 3-8;
PROC SORT; BY donor;

DATA tnfdata42;
INFILE 'a:\data\tnfdata42.dat'
  TRUNCOVER;
INPUT donor 1-2 rep2 3-8;
PROC SORT; BY donor;

DATA tnfdata43;
INFILE 'a:\data\tnfdata43.dat'
  TRUNCOVER;
INPUT donor 1-2 rep3 3-8;
PROC SORT; BY donor;

DATA tnfdata4;
MERGE tnfdata41 tnfdata42 tnfdata43;
BY donor;
mtb=1; fat=1;

DATA abcxyz.tnfdata;
SET tnfdata1 tnfdata2 tnfdata3 tnfdata4;
LABEL rep1='1st replicate';
LABEL rep2='2nd replicate';
LABEL rep3='3rd replicate';
LABEL mtb='MTB presence';
LABEL fat='FAT presence';
tnfaver=(rep1+rep2+rep3)/3;
LABEL tnfaver='average TNF';
FORMAT mtb mtblab.; FORMAT fat fatlab.;

PROC PRINT DATA=abcxyz.tnfdata;

RUN;
```

SAS Output for Program 4.3.1

Obs	donor	rep1	rep2	rep3	mtb	fat	tnfaver
1	1	-0.01	-0.01	-0.13	No MTB	No FAT	-0.05
2	2	16.13	-9.62	-14.88	No MTB	No FAT	-2.79
3	3	.	-0.30	-0.95	No MTB	No FAT	.
4	4	3.63	47.50	55.20	No MTB	No FAT	35.44
5	5	-3.21	-5.64	-5.32	No MTB	No FAT	-4.72
6	6	16.26	52.21	17.93	No MTB	No FAT	28.80
7	7	-12.74	-5.23	-4.06	No MTB	No FAT	-7.34
8	8	-4.67	20.10	110.00	No MTB	No FAT	41.81
9	9	-5.40	20.00	10.30	No MTB	No FAT	8.30
10	10	-10.94	-5.26	-2.73	No MTB	No FAT	-6.31
11	11	-4.19	-11.83	-6.29	No MTB	No FAT	-7.44
12	1	-0.05	-0.09	-0.08	MTB present	No FAT	-0.07
13	2	-9.41	-6.30	5.40	MTB present	No FAT	-3.44
14	3	-10.13	-16.48	-14.79	MTB present	No FAT	-13.80
15	4	8.75	134.90	203.70	MTB present	No FAT	115.78
16	5	612.30	898.20	854.20	MTB present	No FAT	788.23
17	6	2034.00	2743.00	2772.00	MTB present	No FAT	2516.33
18	7	978.50	1137.00	850.00	MTB present	No FAT	988.50
19	8	279.30	124.80	222.10	MTB present	No FAT	208.73

20	9	688.10	530.90	720.20	MTB present	No FAT	646.40
21	10	908.80	811.90	746.40	MTB present	No FAT	822.37
22	11	439.30	960.90	593.30	MTB present	No FAT	664.50
23	1	88.16	97.58	66.27	No MTB	FAT present	84.00
24	2	196.50	114.10	134.20	No MTB	FAT present	148.27
25	3	6.021	.	193.38	No MTB	FAT present	3.53
26	4	935.40	1011.00	951.20	No MTB	FAT present	965.87
27	5	606.00	592.70	608.40	No MTB	FAT present	602.37
28	6	1457.00	1349.00	1625.00	No MTB	FAT present	1477.00
29	7	139.70	399.50	91.69	No MTB	FAT present	210.30
30	8	196.70	270.80	160.70	No MTB	FAT present	209.40
31	9	135.20	221.50	268.00	No MTB	FAT present	208.23
32	10	-14.40	79.62	304.10	No MTB	FAT present	123.11

33	11	516.30	585.90	562.60	No MTB	FAT present	554.93
34	1	709.30	874.30	630.00	MTB present	FAT present	737.87
35	2	4541.00	4106.00	4223.00	MTB present	FAT present	4290.00
36	3	391.00	194.00	254.00	MTB present	FAT present	279.67
37	4	2913.00	3632.00	3417.00	MTB present	FAT present	3320.67
38	5	3801.00	3112.00	3681.00	MTB present	FAT present	3531.33
39	6	10150.00	9410.00	9243.00	MTB present	FAT present	9601.00
40	7	6736.00	6323.00	5117.00	MTB present	FAT present	6058.67
41	8	1454.00	2250.00	1092.00	MTB present	FAT present	1598.67
42	9	857.20	1339.00	1945.00	MTB present	FAT present	1380.40
43	10	.	739.90	4379.00	MTB present	FAT present	.
44	11	6637.00	6909.00	6453.00	MTB present	FAT present	6666.33

4.4 FURTHER MANIPULATION OF DATASETS

The permanent dataset for all the TNF data created above can be further manipulated in order to carry out some basic analyses. In Program 4.4.1, a new variable, "tnfgrp", is created based on the values of "tnfaver". Redundant variables are also removed from the dataset. The updated dataset is then printed out: first, just those cases with a "tnfgrp" value of 1, then, each of the three created groups, as defined by "tnfgrp".

PROGRAM 4.4.1

```
Comments omitted
LIBNAME abcxyz 'a:\data';
LIBNAME library 'a:\formats';

DATA tnf;
SET abcxyz.tnfdata;
DROP rep1 rep2 rep3;
The DROP statement deletes the listed variables from the
dataset "tnf".

IF tnfaver>=988.5 THEN tnfgrp=1;
  ELSE IF tnfaver<=8.3 THEN tnfgrp=3;
     ELSE tnfgrp=2;
The IF statement together with the ELSE statements are now
used to define the groups consisting of the top 25% TNF
averages (Group 1) and the bottom 25% TNF averages
(Group 3), with the rest in Group 2. A new variable "tnfgrp"
that contains the group number for each case is created.

PROC PRINT DATA=tnf;
WHERE tnfgrp=1;
The WHERE statement requests that the PROC PRINT only
operate on those cases in the dataset that have "tnfgrp"
values equal to 1.

PROC SORT DATA=tnf;
BY tnfgrp;
```

```
PROC PRINT DATA=tnf;
BY tnfgrp;
```
Once the dataset is sorted by the "tnfgrp" variable, the BY statement can be used to perform the PROC PRINT repeatedly for each value of "tnfgrp".
```
RUN;
```

SAS Output for Program 4.4.1

Obs	mtb	fat	donor	tnfaver	tnfgrp
17	No MTB	FAT present	6	1477.00	1
28	MTB present	No FAT	6	2516.33	1
29	MTB present	No FAT	7	988.50	1
35	MTB present	FAT present	2	4290.00	1
37	MTB present	FAT present	4	3320.67	1
38	MTB present	FAT present	5	3531.33	1
39	MTB present	FAT present	6	9601.00	1
40	MTB present	FAT present	7	6058.67	1
41	MTB present	FAT present	8	1598.67	1
42	MTB present	FAT present	9	1380.40	1
44	MTB present	FAT present	11	6666.33	1

-----------------------tnfgrp = 1-----------------------

Obs	mtb	fat	donor	tnfaver
1	No MTB	FAT present	6	1477.00
2	MTB present	No FAT	6	2516.33
3	MTB present	No FAT	7	988.50
4	MTB present	FAT present	2	4290.00
5	MTB present	FAT present	4	3320.67
6	MTB present	FATpres ent	5	3531.33
7	MTB present	FAT present	6	9601.00
8	MTB present	FAT present	7	6058.67
9	MTB present	FAT present	8	1598.67
10	MTB present	FAT present	9	1380.40
11	MTB present	FAT present	11	6666.33

---------------------tnfgrp = 2---------------------

Obs	mtb	fat	donor	tnfaver
12	No MTB	No FAT	4	35.443
13	No MTB	No FAT	6	28.800
14	No MTB	No FAT	8	41.810
15	No MTB	FAT present	1	84.003
16	No MTB	FAT present	2	148.267

17	No MTB	FAT present	4	965.867
18	No MTB	FAT present	5	602.367
19	No MTB	FAT present	7	210.297
20	No MTB	FAT present	8	209.400
21	No MTB	FAT present	9	208.233
22	No MTB	FAT present	10	123.083
23	No MTB	FAT present	11	554.933
24	MTB present	No FAT	4	115.783
25	MTB present	No FAT	5	788.233
26	MTB present	No FAT	8	208.733
27	MTB present	No FAT	9	646.400
28	MTB present	No FAT	10	822.367
29	MTB present	No FAT	11	664.500
30	MTB present	FAT present	1	737.867
31	MTB present	FAT present	3	279.667

```
-------------------tnfgrp = 3--------------------
```

| Obs | mtb | fat | donor | tnfaver |
| 32 | No MTB | No FAT | 1 | -0.05000 |

33	No MTB	No FAT	2	−2.76333
34	No MTB	No FAT	3	.
35	No MTB	No FAT	5	−4.72333
36	No MTB	No FAT	7	−7.34333
37	No MTB	No FAT	9	8.30000
38	No MTB	No FAT	10	−6.31000
39	No MTB	No FAT	11	−7.43667
40	No MTB	FAT present	3	3.5300
41	MTB present	No FAT	1	−0.0733
42	MTB present	No FAT	2	−3.4367
43	MTB present	No FAT	3	−13.7700
44	MTB present	FAT present	10	.

Additional Notes for Program 4.4.1

DROP rep1 rep2 rep3; — Redundant variables are removed from the dataset "tnf" using the DROP statement. An alternative would have been to use the KEEP statement: keep mtb fat donor tnfaver;.

IF tnfaver>=988.5 THEN tnfgrp=1; ELSE IF tnfaver<=8.3 THEN tnfgrp=3; ELSE tnf-grp=2; —This series of logical statements defines the new "tnfgrp" variable, based on the "tnfaver" variable. This code can be translated into spoken or written English, as follows: If the variable "tnfaver" (IF tnfaver) is greater than or equal to the value 988.5 (> =988.5), then make the variable "tnf-grp" be equal to 1 (THEN tnfgrp = 1). If this is not true (ELSE), then if the variable "tnfaver" (IF tnfaver) is less than or equal to 8.3 (< =8.3), then make the variable "tnfgrp" be equal to 3 (THEN tnfgrp = 3). If this also is not true (ELSE), then make the variable "tnfgrp" be equal to 2 (tnfgrp = 2).

Other comparison operators such as ">=" (greater than or equal to) and "<=" (less than or equal to) include ">" (strictly greater than), "<" (strictly less than), "=" (equals), and "^=" (not equal to).

PROC PRINT DATA=tnf; WHERE tnfgrp=1; — Use of the WHERE statement is a powerful way of asking SAS to carry out procedures on only part of a dataset. The procedure uses just those cases the WHERE statement requests, and the dataset is not altered by the operation of the WHERE statement.

PROC SORT DATA=tnf; BY tnfgrp; PROC PRINT DATA=tnf; BY tnfgrp; — When it is necessary for a procedure to operate separately on different groups in a dataset, the BY statement can be used in conjunction with the procedure. Here, PROC PRINT is operated three times: once for the cases where "tnfgrp" is 1, again for the cases where "tnfgrp" is 2, and finally for the cases where "tnfgrp" is 3.

4.5 TASKS

These tasks refer to the Nurses' Glove Use Data in Appendix A.

Task 4.1

The "gloves.dat" data file is the result of merging two files: "glovesobs.dat" and "glovesexp.dat". The former contains the observational data, and the latter contains the number of years experience for each nurse. Construct a program that reads in both data files and merges them to produce a dataset in the format of the "gloves.dat" data file previously used. Use PROC PRINT to check the merged dataset.

TASK **4.2**

1. Experiment with the program from Task 4.1 by changing the order in which the datasets are listed in the MERGE command.
2. Using "(IN = variable name)" in the MERGE command, experiment with logical statements such as "IF varname1;".

TASK **4.3**

Using IF statements, create a new variable that indicates whether or not a nurse has eight or more years of experience. Use PROC PRINT to check the results. You may wish to access the data by reading it from the plain text file or from the permanent dataset created in Task 3.1.

5 Restructuring Datasets

5.1 SPLITTING ONE ROW OF DATA INTO SEVERAL ROWS

Use of the data step in Chapter 2 has been limited to reading in text files already laid out in the format required, creating a new variable, and dropping or keeping variables. However, within the data step, a lot can be done to the data to prepare it for analyses with procedures that follow. This can be particularly useful if the data file does not have the required structure.

Suppose that the TNF data came in the structure shown in Table 5.1, with all three replicates alongside each other and two sets of replicates on one line.

Program 5.1.1 reads in this dataset and restructures it so that there is one line in the dataset per TNF reading, as in Table 5.2. It saves the resulting dataset for use in Program 5.2.1.

PROGRAM 5.1.1

```
Comments omitted
LIBNAME abcxyz 'a:\data';
DATA abcxyz.tnfdatasplit;
INFILE 'a:\data\tnfdataalongside.dat'
   TRUNCOVER;
ARRAY abc[12] x1-x12;
```
An array called "abc" is defined to contain 12 variables by having the "[12]". Any 12 names could have been used after "abc[12]", but a shortcut is used here. By using the "x1-x12", the variables "x1", "x2", "x3", "x4", "x5", "x6", "x7", "x8", "x9", "x10", "x11", and "x12" were created.

TABLE 5.1
TNF Data before Rows Split

MTB	FAT	Donor	Rep1	Rep2	Rep3
0	0	1	-0.01	-0.01	-0.13
0	0	2	16.13	-9.62	-14.88
0	0	3	Missing	-0.3	-0.95
0	0	4	3.63	47.5	55.2
0	0	5	-3.21	-5.64	-5.32
0	0	6	16.26	52.21	17.93
0	0	7	-12.74	-5.23	-4.06
0	0	8	-4.67	20.1	110
0	0	9	-5.4	20	10.3
0	0	10	-10.94	-5.26	-2.73
0	0	11	-4.19	-11.83	-6.29
0	1	1	88.16	97.58	66.27
0	1	2	196.5	114.1	134.2
0	1	3	6.02	1.19	3.38
0	1	4	935.4	1011	951.2
0	1	5	606	592.7	608.4
0	1	6	1457	1349	1625
0	1	7	139.7	399.5	91.69
0	1	8	196.7	270.8	160.7
0	1	9	135.2	221.5	268
0	1	10	-14.47	79.62	304.1
0	1	11	516.3	585.9	562.6

MTB	FAT	Donor	Rep1	Rep2	Rep3
1	0	1	-0.05	-0.09	-0.08
1	0	2	-9.41	-6.3	5.4
1	0	3	-10.13	-16.48	-14.79
1	0	4	8.75	134.9	203.7
1	0	5	612.3	898.2	854.2
1	0	6	2034	2743	2772
1	0	7	978.5	1137	850
1	0	8	279.3	124.8	222.1
1	0	9	688.1	530.9	720.2
1	0	10	908.8	811.9	746.4
1	0	11	439.3	960.9	593.3
1	1	1	709.3	874.3	630
1	1	2	4541	4106	4223
1	1	3	391	194	254
1	1	4	2913	3632	3417
1	1	5	3801	3112	3681
1	1	6	10150	9410	9243
1	1	7	6736	6323	5117
1	1	8	1454	2250	1092
1	1	9	857.2	1339	1945
1	1	10	Missing	739.9	4379
1	1	11	6637	6909	6453

```
INPUT x1 1 x2 2 x3 3-4 x4 5-10 x5 11-16
   x6 17-22 x7 23 x8 24 x925-26 x10 27-32
   x11 33-38 x12 39-44;
mtb = x1; fat = x2; donor = x3;
DO i = 4 TO 6;
 tnf = abc[i];
 replicate = i-3;
 OUTPUT;
END;
mtb = x7; fat = x8; donor = x9;
DO i = 10 TO 12;
 tnf = abc[i];
 replicate = i-9;
 OUTPUT;
END;
DROP x1-x12 i;
PROC PRINT DATA = abcxyz.tnfdatasplit;
RUN;
```

SAS Output for Program 5.1.1

Obs	mtb	fat	donor	tnf	replicate
1	0	0	1	−0.01	1
2	0	0	1	−0.01	2
3	0	0	1	−0.13	3
4	1	0	1	−0.05	1
5	1	0	1	−0.09	2
6	1	0	1	−0.08	3
7	0	0	2	16.13	1
.	.	.	.	.	.
.	.	.	.	.	.
.	.	.	.	.	.

126	1	1	10	4379.00	3
127	0	1	11	516.30	1
128	0	1	11	585.90	2
129	0	1	11	562.60	3
130	1	1	11	6637.00	1
131	1	1	11	6909.00	2
132	1	1	11	6453.00	3

Additional Notes for Program 5.1.1

ARRAY abc[12] x1-x12; — This sets up a one-dimensional array called "abc" that has 12 elements. The variables "x1" to "x12" are assigned to these 12 elements. This does not create more variables. The "x1" to "x12" still exist as ordinary variables, but having them defined as elements of an array simply means that they can be referred to as "abc[1]" for "x1", "abc[2]" for "x2", ..., "abc[12]" for "x12". This flexibility is exploited in the "do loops" that follow in the program.

mtb = x1; fat = x2; donor = x3; DO i = 4 TO 6; tnf = abc[i]; replicate = i-3; OUTPUT; END; — SAS effectively executes the transformations in the data step once for each row of data it read in. Thus, for the first row of data, it defines new variables "mtb" to be equal to the row's value for "x1", "fat" to be equal to its value for "x2", and "donor" to be equal to its value for "x3". That is, "mtb = 0", "fat = 0", and "donor = 1".

It then enters a loop. For i = 4, it makes the new variable "tnf" equal to the fourth element of the "abc" array (tnf = abc[i];). It also creates an index variable. It then uses OUTPUT to write all the variables currently stored

TABLE 5.2
TNF Data after Rows Split

MTB	FAT	Donor	Replicate	TNF
0	0	1	1	-0.01
0	0	1	2	-0.01
0	0	1	3	-0.13
1	0	1	1	-0.05
1	0	1	2	-0.09
1	0	1	3	-0.08
0	0	2	1	16.13
.	.	.	.	.
.	.	.	.	.
.	.	.	.	.
1	1	10	3	4379
0	1	11	1	516.3
0	1	11	2	585.9
0	1	11	3	562.6
1	1	11	1	6637
1	1	11	2	6909
1	1	11	3	6453

("x1" to "x12"; the recently declared variables "mtb", "fat", "donor", "tnf", and "replicate"; and the loop indicator, "i") to the dataset.

The END statement is reached, and because "i" has not yet reached the end of its loop, it proceeds to i = 5, and new values for "tnf" and "replicate" are declared. The previously declared "x1" to "x12", "mtb", "fat", and "donor", together with the new "tnf" and "rep" and the loop counter "i", are output.

Next, the same occurs for i = 6, and then finally, the loop is exhausted.

TABLE 5.3
TNF Data after Rows Combined

MTB	FAT	Donor	Rep1	Rep2	Rep3
0	0	1	-0.01	-0.01	-0.13
0	0	2	16.13	-9.62	-14.88
0	0	3		-0.3	-0.95
0	0	4	3.63	47.5	55.2
.	.	.	.	.	.
.	.	.	.	.	.
.	.	.	.	.	.
1	1	8	1454	2250	1092
1	1	9	857.2	1339	1945
1	1	10		739.9	4379
1	1	11	6637	6909	6453

The data step then returns to the beginning and carries out the same procedure for the second row in the dataset.

```
mtb = x7; fat = x8; donor = x9; DO i =
10 TO 12; tnf = abc[i]; replicate = i-
9; OUTPUT; END; — The above process is repeated for
```
"x7" to "x12".

```
DROP x1-x12 i; — Finally, before the PROC PRINT is
```
used, the now unwanted variables "x1" to "x12" and "i" are dropped from the dataset.

5.2 COMBINING SEVERAL ROWS OF DATA INTO ONE ROW

In this section, the TNF data in the structure found in Table 5.2 are restructured to the data shown in Table 5.3, with the three replicates all in one row. Program 5.2.1 contains the necessary code.

PROGRAM 5.2.1

Comments omitted

```
LIBNAME abcxyz 'a:\data';
DATA tnf;
SET abcxyz.tnfdatasplit;
PROC SORT DATA = tnf;
BY mtb fat donor;
```
The data are accessed in the format created above, and PROC SORT is used to sort the data by "mtb", "fat", then "donor". This sets up the possibility to use "LAST.donor" below. A "FIRST.donor" variable is also possible, and also FIRST and LAST for "mtb" and "fat", but these are not needed.

```
DATA tnf; SET tnf;
BY mtb fat donor;
```
Opening the data step again with the BY statement sets up the "LAST.donor" variable.

```
ARRAY reparray[3] rep1-rep3;
RETAIN rep1-rep3;
```
An array is created to contain the replicates in one row, and the RETAIN statement is used so that "rep1" is not lost when "rep2" and "rep3" are created, and "rep2" is not lost when "rep3" is created.

```
reparray[replicate] = tnf;
```
The array is defined with the "tnf" variable. By using the array, "rep1" gets assigned the "tnf" value associated with "replicate" being equal to 1, and similarly for "rep2" and "rep3".

```
IF LAST.donor THEN OUTPUT;
```
Just before SAS moves on to the next donor in the list (i.e., at the position "LAST.donor"), the row of data is output.

```
DROP tnf replicate;
```
The "tnf" and "replicate" variables are no longer required.
```
PROC PRINT DATA = tnf;
RUN;
```

SAS Output for Program 5.2.1

Obs	mtb	fat	donor	rep1	rep2	rep3
1	0	0	1	-0.01	-0.01	-0.13
2	0	0	2	16.13	-9.62	-14.88
3	0	0	3	.	-0.30	-0.95
4	0	0	4	3.63	47.50	55.20
5	0	0	5	-3.21	-5.64	-5.32
6	0	0	6	16.26	52.21	17.93
7	0	0	7	-12.74	-5.23	-4.06
8	0	0	8	-4.67	20.10	110.00
9	0	0	9	-5.40	20.00	10.30
10	0	0	10	-10.94	-5.26	-2.73
11	0	0	11	-4.19	-11.83	-6.29
12	0	1	1	88.16	97.58	66.27
13	0	1	2	196.50	114.10	134.20
14	0	1	3	6.02	1.19	3.38
15	0	1	4	935.40	1011.00	951.20
16	0	1	5	606.00	592.70	608.40
17	0	1	6	1457.00	1349.00	1625.00
18	0	1	7	139.70	399.50	91.69
19	0	1	8	196.70	270.80	160.70
20	0	1	9	135.20	221.50	268.00
21	0	1	10	-14.47	79.62	304.10
22	0	1	11	516.30	585.90	562.60
23	1	0	1	-0.05	-0.09	-0.08
24	1	0	2	-9.41	-6.30	5.40

25	1	0	3	-10.13	-16.48	-14.79
26	1	0	4	8.75	134.90	203.70
27	1	0	5	612.30	898.20	854.20
28	1	0	6	2034.00	2743.00	2772.00
29	1	0	7	978.50	1137.00	850.00
30	1	0	8	279.30	124.80	222.10
31	1	0	9	688.10	530.90	720.20
32	1	0	10	908.80	811.90	746.40
33	1	0	11	439.30	960.90	593.30
34	1	1	1	709.30	874.30	630.00
35	1	1	2	4541.00	4106.00	4223.00
36	1	1	3	391.00	194.00	254.00
37	1	1	4	2913.00	3632.00	3417.00
38	1	1	5	3801.00	3112.00	3681.00
39	1	1	6	10150.00	9410.00	9243.00
40	1	1	7	6736.00	6323.00	5117.00
41	1	1	8	1454.00	2250.00	1092.00
42	1	1	9	857.20	1339.00	1945.00
43	1	1	10	.	739.90	4379.00
44	1	1	11	6637.00	6909.00	6453.00

Additional Notes for Program 5.2.1

FIRST.donor, LAST.donor — The PROC SORT; BY mtb
fat donor; sets up the possibility for these two additional
variables to be used in the subsequent data step. They can
be thought of as having the values "true" and "false." In the
sorted dataset, each occurrence of a new donor value is
associated with the "true" value for the "FIRST.donor" variable, and each final occurrence of a donor value is associated
with the "true" value for the "LAST.donor" variable.
Although not used in Program 5.2.1, the variables
"FIRST.mtb", "LAST.mtb", "FIRST.fat", and "LAST.fat"
are also created, because they are included in the BY state-

TABLE 5.4
TNF Data with FIRST.donnor and LAST donor

MTB	FAT	Donor	TNF	Replicate	FIRST. donor	LAST. donor
0	0	1	-0.01	1	True	False
0	0	1	-0.01	2	False	False
0	0	1	-0.13	3	False	True
0	0	2	16.13	1	True	False
.	.	.	.	.	.	.
.	.	.	.	.	.	.
.	.	.	.	.	.	.
1	1	10	4379	3	False	True
1	1	11	6637	1	True	False
1	1	11	6909	2	False	False
1	1	11	6453	3	False	True

ment associated with the data step. Shown in Table 5.4 is what the sorted TNF dataset from Table 5.2 would look like with the additional "FIRST.donor" and "LAST.donor" variables.

ARRAY reparray[3] rep1-rep3; RETAIN rep1-rep3; reparray[replicate] = tnf; IF LAST.donor THEN OUTPUT; — Because the variables "rep1", "rep2", and "rep3" do not yet exist in the dataset, the ARRAY statement creates them and allows them to be referred to as the elements of the "reparray" array.

SAS starts with the first line of the dataset (as shown in Table 5.4, with the additional "FIRST.donor" and "LAST.donor" variables). The value of "replicate" is 1, and the value of "tnf" is -0.01. Thus, the "reparray[replicate] = tnf" statement is effectively "rep1 = -0.01". The value of "LAST.donor" is "false", so the IF statement does not perform the OUTPUT.

As in Program 5.1.1, SAS executes the data step once for each row of data in the dataset, so SAS now moves to the second line of the dataset. The value of "replicate" is 2, and the value of "tnf" is -0.01. Thus, the "reparray[replicate] = tnf" statement is effectively "rep2 = -0.01". The value of "LAST.donor" is "false", so the IF statement does not perform the OUTPUT.

Moving to the third row of the dataset, the value of "replicate" is 3, and the value of "tnf" is -0.13. Thus, the "reparray[replicate] = tnf" statement is effectively "rep3 = -0.13". The value of "LAST.donor" is now "true," so the IF statement performs the OUTPUT. What SAS has available to output at this point is all the values for the third line of the dataset together with "rep3", which was just defined, and also "rep1" and "rep2", which are still available because they were specified in the RETAIN statement. Thus, the variables output to the new dataset being created are as below:

mtb	fat	donor	tnf	replicate	rep1	rep2	rep3
0	0	1	-0.13	3	-0.01	-0.01	-0.13

If the RETAIN statement had not been used, SAS would have forgotton the values of "rep1" and "rep2" (set them to be missing values) before it got to perform the OUTPUT statement.

SAS then moves to Lines 4, 5, and 6 of the dataset and outputs a line, with the same variables as above, for Donor 2, with MTB and FAT values of 0. This continues until SAS processes all the lines of the dataset and a line is output for each combination of "mtb", "fat", and "donor".

DROP tnf replicate; — Finally, the variables "tnf" and
"replicate", which are no longer relevant, are dropped from
the new dataset.

5.3   TASKS

These tasks refer to the Nurses' Glove Use Data in Appendix A.

Task 5.1

The file "glovesonerow.dat" contains data with one row per
nurse. The first variable in this file is the nurse's number; the
next four variables are the numbers of times the nurse was
observed before the educational programme and at 1, 2, and 5
months after the programme; the next four variables are the
numbers of times that the nurse used gloves in each of these
observation periods; the last variable is the number of years of
experience the nurse has in her position.

Create an SAS program to read this data and restructure it so
that there are four rows for each nurse, corresponding to each
of the four periods. Each row should contain the nurse's number,
the period (coded 1, 2, 3, and 4), the number of times the nurse
was observed in the period, the number of times the nurse used
gloves in the period, and the number of years of experience that
the nurse has in her position.

Save the restructured dataset as a permanent dataset, and use PROC PRINT to check the results of the restructuring. It should give a dataset that has the same structure as the plain text file "gloves.dat".

Task 5.2

Taking the permanent dataset created in Task 5.1, carry out a further restructuring that returns the data to the format found in the plain text file "glovesonerow.txt". That is, there should be one row per nurse and the variables "nurse", "obs1", "obs2", "obs3", "obs4", "gloves1", "gloves2", "gloves3", "gloves4", and "experience" should be included. Use PROC PRINT to check the reconstructed dataset.

6 Summarising Data

In this section, the use of PROC MEANS and PROC UNIVARIATE to summarise continuous data and the use of PROC FREQ to summarise categorical data are shown. The PROC FREQ program also introduces the use of the TITLE statement.

6.1 PROC MEANS

PROC MEANS does more than simply calculate means. It is capable of providing a range of summary statistics for variables.

PROGRAM 6.1.1

```
Comments omitted
LIBNAME abcxyz 'a:\data';
LIBNAME library 'a:\formats';

PROC MEANS DATA = abcxyz.tnfdata N MEAN
   STD MIN MAX RANGE;
VAR rep1 rep2 rep3 tnfaver;
```
PROC MEANS is used to produce univariate summaries for each variable in the list, as defined by the VAR statement for the dataset "tnfdata". The keywords N, MEAN, STD, MIN, MAX, and RANGE tell SAS to display the number of observations, mean, standard deviation, minimum, maximum, and range, respectively, for the data. If these keywords were omitted, a default selection would have been displayed.

```
RUN;
```

SAS Output for Program 6.1.1

The MEANS Procedure

Variable	Label	N	Mean	Std Dev	Minimum
rep1	1st replicate	42	1151.82	2178.50	−14.4700000
rep2	2nd replicate	44	1159.84	2048.19	−16.4800000
rep3	3rd replicate	44	1189.12	2010.66	−14.8000000
tnfaver	average TNF	42	1163.59	2077.37	−13.7700000

Variable	Label	Maximum	Range
rep1	1st replicate	10150.00	10164.47
rep2	2nd replicate	9410.00	9426.48
rep3	3rd replicate	9243.00	9257.80
tnfaver	average TNF	9601.00	9614.77

Additional Notes for Program 6.1.1

Operators — In Program 2.5.1, the "+" and "/" symbols were used to calculate average TNF. Other valid arithmetic operators are "-" for subtraction, "*" for multiplication, and "**"

for exponentiation. The Boolean operators AND, OR, and
NOT are also valid.

VAR rep1 rep2 rep3 tnfaver; — If this statement
   had been omitted, SAS would have created summaries for
   all the variables in the "tnfdata" dataset. This would not have
   been appropriate, as, for instance, no meaning can be
   attached to the mean donor number.

Even in situations where summaries of all the variables in the
   dataset are required, it is good practice to use the VAR
   statement so that it is clear to other readers of the program
   what it is meant to be doing.

6.2   PROC UNIVARIATE

PROC UNIVARIATE produces summary statistics, like PROC
MEANS, but it can also produce some useful plots.

PROGRAM 6.2.1

Comments omitted

```
LIBNAME  abcxyz  'a:\data';
LIBNAME  library  'a:\formats';

PROC UNIVARIATE DATA = abcxyz.tnfdata
   PLOTS;
VAR tnfaver;
```

*PROC UNIVARIATE produces a range of statistics. The
PLOTS option means that SAS will produce crude stem-and-
leaf plots, boxplots, and normal probability plots. The use
of further statements with the PROC UNIVARIATE can produce high-resolution graphics.*

```
RUN;
```

SAS Output for Program 6.2.1

```
                  The UNIVARIATE Procedure
             Variable: tnfaver (average TNF)

                          Moments

N                         42    Sum Weights              42
Mean               1163.5923    Sum
                                Observations     48870.8767
Std Deviation     2077.37044    Variance         4315467.93
Skewness          2.59864223    Kurtosis         6.99472746
Uncorrected SS     233799961    Corrected SS      176934185
Coeff Variation                 Std Error Mean
                  178.530782                     320.545217


                Basic Statistical Measures

        Location              Variability

 Mean      1163.592       Std Deviation               2077
 Median     209.848       Variance                 4315468
 Mode          .          Range                       9615
                          Interquartile
                           Range                 980.20000


               Tests for Location: Mu0 = 0

 Test            -Statistic-      - - -p Value - - -

 Student's t   t    3.630041      Pr > |t|      0.0008
 Sign          M          12      Pr > = |M|    0.0003
 Signed Rank   S       400.5      Pr > = |S|    <.0001
```

```
                  Quantiles (Definition 5)

          Quantile                Estimate

          100%  Max               9601.00000
          99%                      9601.00000
          95%                      6058.66667
          90%                      3531.33333
          75%  Q3                   988.50000
          50%  Median               209.84833
          25%  Q1                     8.30000
          10%                        -4.72333
          5%                         -7.34333
          1%                        -13.77000
          0%  Min                  -13.77000



                  Extreme Observations

    - - - Lowest - - -          - - -Highest - - -

     Value          Obs          Value         Obs

   -13.77000         25        3531.33          38
    -7.43667         11        4290.00          35
    -7.34333          7        6058.67          40
    -6.31000         10        6666.33          44
    -4.72333          5        9601.00          39



                    Missing Values

                              - - -Percent Of - - -
  Missing                                  Missing
   Value            Count      All Obs       Obs

     .                 2         4.55       100.00
```

```
Stem       Leaf                          #      Boxplot
   9  6                                   1         *
   9
   8
   8
   7
   7
   6  7                                   1         *
   6  1                                   1         *
   5
   5
   4
   4  3                                   1         *
   3  5                                   1         0
   3  3                                   1         0
   2  5                                   1         0
   2
   1  56                                  2         |
   1  004                                 3      + - - + -
                                                   - +
   0  6667788                            7
   0  00000111122223                    14      * - - - -
                                                   - *
  -0  000000000                          9         |
 - - + - - + - - + - - +
          Multiply Stem.Leaf by 10**+3
```

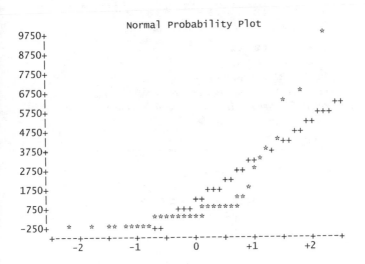

6.3 PROC FREQ

The procedure PROC FREQ can be used to produce frequency tables and cross-tabulations.

PROGRAM 6.3.1

```
Comments omitted
LIBNAME abcxyz 'a:\data';
LIBNAME library 'a:\formats';
PROC FREQ DATA = abcxyz.tnfdata;
```

```
TITLE 'MTB and FAT presence';
```
This statement adds a title to the SAS output.

```
TABLES mtb fat;
TABLES mtb*fat;
```
PROC FREQ is used (with a title). The first TABLES statement produces frequency counts for "mtb" and "fat" separately. The second TABLES statement produces a cross-tabulation of the two variables.

```
RUN;
```

SAS Output for Program 6.3.1

MTB and FAT Presence
The FREQ Procedure

MTB presence

mtb	Frequency	Cumulative Percent	Cumulative Frequency	Percent
No MTB	22	50.00	22	50.00
MTB present	22	50.00	44	100.00

FAT presence

fat	Frequency	Cumulative Percent	Cumulative Frequency	Percent
No FAT	22	50.00	22	50.00
FAT present	22	50.00	44	100.00

```
                MTB  and  FAT  Presence
                   The  FREQ  Procedure
                   Table of mtb by fat
```

mtb(MTB presence)   fat(FAT presence)

Frequency Percent Row Pct Col Pct	No FAT	FAT present	Total
No MTB	11 25.00 50.00 50.00	11 25.00 50.00 50.00	22 50.00
MTB present	11 25.00 50.00 50.00	11 25.00 50.00 50.00	22 50.00
Total	22 50.00	22 50.00	44 100.00

6.4   TASKS

These tasks refer to the Nurses' Glove Use Data in Appendix A.

TASK 6.1

Use PROC MEANS to produce a table that shows summary statistics (mean, 95% limits, minimum, median, maximum) for the proportion of observations in which gloves were used (created in Task 2.1). To find out what keywords to use for the specified summary statistics, use the help facility or refer to documentation.

Task 6.2

Use PROC UNIVARIATE to examine the variable (created in Task 2.1) relating to the proportion of observations in which gloves were used. Use the help facility or refer to documentation to produce a high-resolution normal probability plot.

Task 6.3

Produce a frequency table for the period variable.

Task 6.4

Produce a cross-tabulation of the nurses' identification numbers against periods using PROC FREQ. This should give a table of zeros (where data is missing) and ones (where data is present).

7   Tabulating Data

Simple tables summarising data were produced previously. SAS also has the ability to produce much more complicated tables using PROC MEANS and, especially, using PROC TABULATE.

7.1   PROC MEANS

Shown in Program 7.1.1 is an aspect of PROC MEANS that was not utilised in Section 6.1 — the use of the CLASS statement that causes the summary statistics to be produced for combinations of levels of discrete variables.

PROGRAM 7.1.1

Comments omitted

```
LIBNAME abcxyz 'a:\data';
LIBNAME library 'a:\formats';
PROC MEANS DATA = abcxyz.tnfdata MEAN CLM
   MAXDEC = 2;
VAR rep1 rep2 rep3 tnfaver;
CLASS mtb fat;
```
This more advanced use of PROC MEANS tells SAS to summarise the variables defined by the VAR statement and to show the output broken down by the different classes of "mtb" and "fat". The MEAN option tells SAS to calculate and display the means, and the CLM option tells SAS to compute 95% confidence intervals. The MAXDEC = 2 option tells SAS that the maximum number of decimal places to print is two.
```
RUN;
```

SAS Output for Program 7.1.1

The MEANS Procedure

MTB presence	FAT presence	N Obs	Variable	Label	Mean
No MTB	No FAT	11	rep1	1st replicate	-0.51
			rep2	2nd replicate	9.27
			rep3	3rd replicate	14.47
			tnfaver	average TNF	8.57
	FAT present	11	rep1	1st replicate	387.50
			rep2	2nd replicate	429.35
			rep3	3rd replicate	434.14
			tnfaver	average TNF	417.00
MTB present	No FAT	11	rep1	1st replicate	539.04
			rep2	2nd replicate	665.34
			rep3	3rd replicate	632.05
			tnfaver	average TNF	612.14

MTB presence	FAT presence	N Obs	Variable	Label	Lower 95% CL for Mean
	FAT present	11	rep1	1st replicate	3818.95
			rep2	2nd replicate	3535.38
			rep3	3rd replicate	3675.82
			tnfaver	average TNF	3746.46
No MTB	No FAT	11	rep1	1st replicate	-7.65
			rep2	2nd replicate	-6.00
			rep3	3rd replicate	-10.29
			tnfaver	average TNF	-5.22
	FAT present	11	rep1	1st replicate	79.42
			rep2	2nd replicate	142.78
			rep3	3rd replicate	106.41
			tnfaver	average TNF	112.96
MTB present	No FAT	11	rep1	1st replicate	122.52
			rep2	2nd replicate	117.39
			rep3	3rd replicate	100.48
			tnfaver	average TNF	118.41

	FAT present	11	rep1	1st replicate	1511.24
			rep2	2nd replicate	1557.08
			rep3	3rd replicate	1869.39
			tnfaver	average TNF	1614.32

MTB presence	FAT presence	N Obs	Variable	Label	Upper 95% CL for Mean
No MTB	No FAT	11	rep1	1st replicate	6.62
			rep2	2nd replicate	24.53
			rep3	3rd replicate	39.23
			tnfaver	average TNF	22.37
	FAT present	11	rep1	1st replicate	695.58
			rep2	2nd replicate	715.92
			rep3	3rd replicate	761.87
			tnfaver	average TNF	721.03

MTB present	No FAT	11	rep1	1st replicate	955.56
			rep2	2nd replicate	1213.29
			rep3	3rd replicate	1163.61
			tnfaver	average TNF	1105.88
	FAT present	11	rep1	1st replicate	6126.66
			rep2	2nd replicate	5513.68
			rep3	3rd replicate	5482.25
			tnfaver	average TNF	5878.60

7.2   PROC TABULATE FOR SUMMARISING DATA

Tables such as that produced by PROC MEANS in Program 7.1.1 can also be produced using PROC TABULATE. This is a more flexible tabulating statement that offers the user much more control over the output than that provided by PROC MEANS.

PROGRAM 7.2.1

Comments omitted

```
LIBNAME  abcxyz  'a:\data';
LIBNAME  library  'a:\formats';

PROC  TABULATE  DATA = abcxyz.tnfdata;
CLASS  mtb  fat;
VAR  rep1  rep2  rep3  tnfaver;
TABLE (mtb*fat)*(rep1 rep2 rep3 tnfaver),
   MEAN  STDERR/RTSPACE = 60;
```

Here, the CLASS statement tells SAS which of the variables in the table are categorical variables. The variables to be summarised are defined in the VAR statement. In the TABLES statement, the variables "rep1", "rep2", "rep3", and "tnfaver" are nested within the cross-classification of "mtb" and "fat". As they appear to the left of the comma in the statement, they define the rows. To the right of the comma in the statement, the columns are defined. Here there are only two statistics, MEAN and STDERR, giving the values of mean and standard error of the mean for the variables to be summarised. The RTSPACE = 60 option tells SAS how wide the columns should be.

```
RUN;
```

SAS Output for Program 7.2.1

MTB presence	FAT presence		Mean	StdErr
No MTB	No FAT	1st replicate	-0.51	3.16
		2nd replicate	9.27	6.85
		3rd replicate	14.47	11.11
		average TNF	8.57	6.10
	FAT present	1st replicate	387.50	138.27
		2nd replicate	429.35	128.61
		3rd replicate	434.14	147.09
		average TNF	417.00	136.45

MTB present No FAT			
	1st replicate	539.04	186.94
	2nd replicate	665.34	245.92
	3rd replicate	632.05	238.57
	average TNF	612.14	221.59

FAT present			
	1st replicate	3818.95	1020.14
	2nd replicate	3535.38	887.87
	3rd replicate	3675.82	810.73
	average TNF	3746.46	942.53

7.3   PROC TABULATE FOR CROSS-TABULATIONS

PROC TABULATE can also be used to create cross-tabulations, as in
Program 7.3.1.

PROGRAM 7.3.1

Comments omitted

```
LIBNAME abcxyz 'a:\data';
LIBNAME library 'a:\formats';

PROC TABULATE DATA = abcxyz.tnfdata;
CLASS mtb fat;
TABLE mtb ALL,(fat ALL)*(N*F = 10.0);
KEYLABEL ALL = 'Total' N = 'Number';
```
*Here, the CLASS statement tells SAS which of the variables
in the table are categorical. In the TABLES statement, the
variable "mtb" defines the rows, as it appears to the left of
the comma in the statement. To the right of the comma in
the statement, the columns are defined by the variable "fat",
and the ALL gives the total column. The "fat" variable and
ALL are bracketed together as the (N*F = 10.0) is applied
to them both. This means that the statistic N (number of
observations) will be output with a format of 10.0, which
means it is allowed 10 spaces for display and zero decimal
places.*
```
RUN;
```

SAS Output for Program 7.3.1

	FAT presence		Total
	No FAT	FAT present	
	Number	Number	Number
MTB presence			
No MTB	11	11	22
MTB present	11	11	22
Total	22	22	44

7.4 TASKS

These tasks refer to the Nurses' Glove Use Data in Appendix A.

TASK 7.1

Use PROC MEANS to produce a table that shows summary statistics (mean, 95% limits, minimum, median, maximum) for the proportion of observations where gloves were used (created in Task 2.1), like the one produced for Task 6.1, but this time classified by observation period.

TASK 7.2

Produce a table giving the same information as in Task 7.1 (excluding 95% limits), but this time, use PROC TABULATE.

TASK 7.3

Produce a cross-tabulation of the nurses' identification numbers against periods using PROC TABULATE. This should give a table of ones (where data are present).

8   Formatting and Saving SAS Output

The main means by which output from SAS programs is controlled is with the Output Delivery System, the primary focus of this section. A sometimes useful technique for saving SAS output as a SAS dataset will also be discussed.

8.1   OUTPUT DELIVERY SYSTEM

The Output Delivery System (ODS) allows the user to choose how results of analyses, data manipulations, etc., are presented by SAS.

In Version 8 of SAS, possible destinations for results are:

- LISTING: The "traditional" text output using a monospace font.
- PRINTER: A printer file.
- HTML: Hypertext markup language, as used for Internet pages, with tables of pages and contents using frames, as well as a "body" file containing the results.
- RTF: A "rich text" file that can be read by word processors.
- PDF: A file in "portable document format".
- OUTPUT: A data file containing the data that make up the results.

The various SAS procedures and data steps in a program do not directly produce viewable output. What they actually produce is a set of data. This may take the form of various summary statistics (e.g., for PROC MEANS or PROC UNIVARIATE) or other relevant figures (e.g., cell counts from PROC FREQ). ODS takes this set of data and combines it with a "table definition" that is particular to the operation that created the data. These table definitions are part of the SAS package.

So that the user can see what sort of output has been produced, SAS stores references to all the different bits in a Results folder. In the Microsoft Windows version of SAS, the contents of this Results folder can be seen in a similar way as folders and files are viewed in Microsoft Windows Explorer. SAS uses different icons for different ODS destinations.

8.1.1 LISTING

By default, the LISTING destination is open, and the other destinations are closed. That is, results are presented in the traditional text-only fashion. The LISTING destination can be closed by using the statement ODS LISTING CLOSE; and can be reopened at any time with the ODS LISTING; statement.

In the Microsoft Windows environment, the contents of the LISTING destination can be saved using "File", "Save As...", with the Output window active. This enables the user to save the window's contents as "RTF" using "Save as type:". This file format can be read into a word processor for editing. Sections of the Output window can also be selected in the normal way, using the mouse or cursor keys. The usual copy and paste buttons (or options under "Edit") can be used to transfer the selection to another package.

8.1.2 PRINTER

A printer file can be created to hold the results from SAS. It is closed by default and is initialised by the statement ODS PRINTER FILE = 'filename.ps' PS;. The PS option tells SAS to send results in postscript format to the file 'filename.ps' until the statement ODS PRINTER CLOSE; is used. PDF (portable document format) and PCL (Hewlett Packard's printer control language) are alternatives to postscript.

8.1.3 HTML

The HTML destination is also closed by default. The minimum that must be specified to open it is ODS HTML BODY = 'bodyfile.htm';.

Results will then be put into the "body file" called 'bodyfile.htm' until the statement ODS HTML CLOSE; is used.

It is possible to create three additional HTML files: a table of contents, a table of pages, and a frame in which to display them and the body file. This is done using the following extended statement:

ODS HTMLBODY = 'bodyfile.htm'
CONTENTS = 'contentsfile.htm'
PAGE = 'pagefile.htm'
FRAME = 'framefile.htm';

The statement ODS HTML CLOSE; must be used before the html file can be browsed (which is done using a normal Internet browser). However, if the user wishes to send results to the same body file at different points in a program, then the first time the body file is opened, it must be done using ODS HTML BODY = 'bodyfile.htm' (NO_BOTTOM_MATTER);. Then when ODS HTML CLOSE; is used, the html file is not finished off and is ready to receive more results.

The next time the user wishes to output results to this same body file, the statement that must be used is ODS HTML BODY = 'bodyfile.htm' (NO_TOP_MATTER NO_BOTTOM_MATTER);. Then the normal starting material for an html file is not added to the body file, and again it is not finished off when ODS HTML CLOSE; is used.

The last time the user wishes to output results to this same body file, the statement that must be used is ODS HTML BODY = 'bodyfile.htm' (NO_TOP_MATTER);. Then the normal starting material for an html file is not added to the body file, but the file is properly finished off when ODS HTML CLOSE; is used.

8.1.4   RTF

An RTF file can also be created to hold the results from SAS. It is also closed by default and initialised by the statement ODS RTF FILE = 'filename.rtf';. Results will be sent to the file 'filename.rtf' until the statement ODS RTF CLOSE; is used.

8.1.5   PDF

Another type of file that can be created to hold the results from SAS is
a PDF file. Again, it is closed by default and initialised by the statement
ODS PDF FILE = 'filename.pdf';. Results will be sent to this file until
the statement ODS PDF CLOSE; is used.

8.1.6   OUTPUT

The dataset produced by an SAS procedure or data step can be treated
in the same way as any other SAS dataset. Frequently, more than one
dataset is produced by an SAS procedure, one for each "object" output
by the procedure. These objects have particular names in SAS, and it is
these names that are used in place of "output_object_name" in the
statement below. The "new_dataset_name" is whatever the user wishes
to call the dataset:

```
ODS OUTPUT output_object_name =
    new_dataset_name;
```

It is not necessary to issue a close statement, because once the dataset
is output, the action is completed and needs no closing.

8.1.7   ODS TRACE

To keep track of all the results, the statement ODS TRACE ON; can be
useful. This puts information about output in the log file and continues
until the statement ODS TRACE OFF; is used.

8.1.7.1   ODS [destination] SELECT

Rather than having ODS output all the results that a procedure or data
step produces, particular output objects can be selected for output (using
ODS [destination] SELECT output_object_name;) or excluded from
being output (using ODS [destination] EXCLUDE
output_object_name;). The "[destination]" is replaced by whatever destination is required. The default for the LISTING, PRINTER, and HTML
destinations is SELECT ALL. For the OUTPUT destination, the default
is EXCLUDE ALL.

8.1.8   PROC TEMPLATE

The table definitions used by ODS can be edited using PROC TEM-
PLATE (details not given here). The original definition is not erased.
The new edited version is simply discovered first by SAS and used. New
table definitions can also be created, and SAS can then be told to use
these templates.

PROC TEMPLATE can also be used to change the style of the
output for an ODS destination as a whole or just for elements of output
or even cells within elements of output.

8.1.9   EXAMPLE USING ODS DESTINATIONS, TRACE AND SELECT

The variables "rep1" and "tnfaver" are to be summarised with PROC
UNIVARIATE. However, not all the output produced by PROC
UNIVARIATE is required, and different destinations are required for
different parts of the output:

- The plots for "tnfaver" need to go to the LISTING desti-
  nation.
- The means, variances, etc., for both variables need to go
  to an RTF file via the RTF destination.
- The quantiles need to be output to a dataset by using the
  OUTPUT destination and then need to be printed to the
  HTML destination.

In order to choose the relevant parts of the PROC UNIVARIATE
output using SELECT, it is necessary to know the specific names SAS
gives to the pieces of output. The ODS TRACE statement is used in
Program 8.1.1 to help with this. It produces information in the log file,
which is shown below.

PROGRAM 8.1.1

```
Comments omitted
LIBNAME abcxyz 'a:\data';
```

```
LIBNAME library 'a:\formats';
ODS TRACE ON;
ODS LISTING CLOSE;
```
The ODS TRACE ON statement means that the trace infor-mation will be placed in the log file. The ODS LISTING CLOSE prevents output from the PROC UNIVARIATE to this destination (which is initially open by default), and as no other destination is open, no output will be produced.

```
PROC UNIVARIATE DATA = abcxyz.tnfdata
   PLOTS;
VAR rep1 tnfaver;

RUN;
```

The RUN statement at this point in the program forces SAS to carry out the PROC UNIVARIATE before considering the rest of the program.

```
ODS TRACE OFF;
ODS LISTING;
```
The statement ODS TRACE OFF turns off the recording of trace information. The statement ODS LISTING restores the default open status of this destination.

```
RUN;
```

Part of SAS Log for Program 8.1.1

```
1                          /*
2                          Author: Neil Spencer
3                          Creation Date: ??/??/??
4                          Revision 1 Date: ??/??/??
5                          Purpose: Program 8.1.1
6                          */
7
8                          libname abcxyz 'a:\data';
NOTE: Libref ABCXYZ was successfully assigned as
      follows:
```

```
        Engine:          V8
        Physical Name:   a:\data
9       libname library 'a:\formats';
NOTE: Libref LIBRARY was successfully assigned as
        follows:
        Engine:          V8
        Physical Name:   a:\formats
10
11                      ods traceon;
12                      ods listing close;
13
14                      proc univariate data =
                         abcxyz.tnfdata plots;
15                      var rep1 tnfaver;
16
17                      run;
WARNING: No output destinations active.

Output Added:
- - - - - - -
Name:                   Moments
Label:                  Moments
Template:               base.univariate.Moments
Path:                   Univariate.rep1.Moments
- - - - - - -
Output Added:
- - - - - - -
Name:                   BasicMeasures
Label:                  Basic Measures of Location and
                         Variability
Template:               base.univariate.Measures
Path:                   Univariate.rep1.BasicMeasures
- - - - - - -
Output Added:
- - - - - - -
Name:                   TestsForLocation
```

```
Label:                    Tests For Location
Template:                 base.univariate.Location
Path:                     Univariate.rep1.TestsFor
                            Location

 — — — — — — —
Output Added:
 — — — — — — —

Name:                     Quantiles
Label:                    Quantiles
Template:                 base.univariate.Quantiles
Path:                     Univariate.rep1.Quantiles
 — — — — — — —
Output Added:
 — — — — — — —

Name:                     ExtremeObs
Label:                    Extreme Observations
Template:                 base.univariate.ExtObs
Path:                     Univariate.rep1.ExtremeObs
 — — — — — — —
Output Added:
 — — — — — — —

Name:                     MissingValues
Label:                    Missing Values
Template:                 base.univariate.Missings
Path:                     Univariate.rep1.MissingValues

Output Added:
 — — — — — — —

Name:                     Plots
Label:                    Plots
Data Name:                BatchOutput
Path:                     Univariate.rep1.Plots
 — — — — — — —
Output Added:
 — — — — — — —
```

```
Name:               Moments
Label:              Moments
Template:           base.univariate.Moments
Path:               Univariate.tnfaver.Moments
  - - - - - - -
Output Added:
  - - - - - - -
Name:               BasicMeasures
Label:              Basic Measures of Location and
                     Variability
Template:           base.univariate.Measures
Path:               Univariate.tnfaver.Basic
                     Measures

  - - - - - - -
Output Added:
  - - - - - - -
Name:               TestsForLocation
Label:              Tests For Location
Template:           base.univariate.Location
Path:               Univariate.tnfaver.TestsFor
                     Location

  - - - - - - -
Output Added:
  - - - - - - -
Name:               Quantiles
Label:              Quantiles
Template:           base.univariate.Quantiles
Path:               Univariate.tnfaver.Quantiles
  - - - - - - -
Output Added:
  - - - - - - -
Name:               ExtremeObs
Label:              Extreme Observations
Template:           base.univariate.ExtObs
Path:               Univariate.tnfaver.ExtremeObs
```

```
— — — — — — —
Output Added:
— — — — — — —
Name:                        MissingValues
Label:                       Missing Values
Template:                    base.univariate.Missings
Path:                        Univariate.tnfaver.Missing
                              Values

— — — — — — —
Output Added:
— — — — — — —
Name:                        Plots
Label:                       Plots
Data Name:                   BatchOutput
Path:                        Univariate.tnfaver.Plots
— — — — — — —
```

8.1.10 NAMES REQUIRED

From the trace information in the log file, it can be seen that the PROC
UNIVARIATE output consists of "Moments", "BasicMeasures", "Tests-
ForLocation", "Quantiles", "ExtremeObs", "MissingValues", and
"Plots", first for "rep1" and then for "tnfaver". Thus, to refer to the
means, variances, etc., and the quantiles for both variables, the simple
names "Moments" and "Quantiles" need to be used. To refer to the plots
for the "tnfaver" variable only, the more complex name "Univari-
ate.tnfaver.Plots" needs to be used.

Program 8.1.2 uses these names to produce the output to the
LISTING, RTF, and OUTPUT destinations.

PROGRAM 8.1.2

```
Comments omitted
LIBNAME abcxyz 'a:\data';
LIBNAME library 'a:\formats';
```

```
ODS RTF FILE = 'a:\example.rtf';
ODS RTF SELECT MOMENTS;
```
*The RTF destination is opened, and the file "a:\example.rtf"
is specified. The ODS RTF SELECT statement determines that
just the output named "Moments" will be sent to this destination.*

```
ODS LISTING SELECT
  Univariate.tnfaver.Plots;
```
*The ODS LISTING SELECT statement determines that just
the output named "Univariate.tnfaver.Plots" will be sent to
this destination.*

```
ODS OUTPUT Quantiles =
  abcxyz.quantilesoutput;
```
*The ODS OUTPUT statement determines that the output
named "Quantiles" will be sent to a permanent dataset
called "quantilesoutput".*

```
PROC UNIVARIATE DATA = abcxyz.tnfdata
  PLOTS;
VAR rep1 tnfaver;

RUN;
```
*The RUN statement at this point in the program forces SAS
to carry out the PROC UNIVARIATE before considering the
rest of the program.*

```
ODS RTF CLOSE;
```
*No more output to the RTF destination is required, so the
destination is closed.*

```
RUN;
```

SAS Output for Program 8.1.2 (RTF Destination)

Moments			
N	42	Sum Weights	42
Mean	1151.81738	Sum Observations	48376.33
Std Deviation	2178.49714	Variance	4745849.79
Skewness	2.72371838	Kurtosis	7.6237416
Uncorrected SS	250300539	Corrected SS	194579841
Coeff Variation	189.135637	Std Error Mean	336.149407

Moments			
N	42	Sum Weights	42
Mean	1163.5923	Sum Observations	48870.8767
Std Deviation	2077.37044	Variance	4315467.93
Skewness	2.59864223	Kurtosis	6.99472746
Uncorrected SS	233799961	Corrected SS	176934185
Coeff Variation	178.530782	Std Error Mean	320.545217

SAS Output for Program 8.1.2 (LISTING Destination)

The UNIVARIATE Procedure
Variable: tnfaver (average TNF)

```
             Stem  Leaf                       #

Boxplot       9    6                          1    *
              9
              8
              8
              7
              7
              6    7                          1    *
              6    1                          1    *
              5
              5
              4
              4    3                          1    *
              3    5                          1    0
              3    3                          1    0
              2    5                          1    0
              2
              1    56                         2            |
              1    004                        3    + — + — +
              0    6667788                    7        | |
              0    00000111122223            14    *  — — *
             -0    000000000                  9            |
          — — + — — + — — + — — +
Multiply Stem.Leaf by 10**+3
```

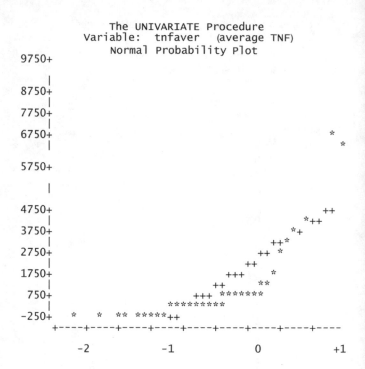

8.1.11 PRINTING THE "QUANTILESOUTPUT" DATASET TO THE HTML DESTINATION

With the "quantilesoutput" stored as a permanent dataset, Program 8.1.3 uses PROC PRINT to send it to the HTML destination.

PROGRAM 8.1.3

Comments omitted

```
LIBNAME abcxyz 'a:\data';
LIBNAME library 'a:\formats';
```

```
ODS LISTING CLOSE;
```
The ODS LISTING CLOSE prevents output from the PROC PRINT from coming to this destination.

```
ODS HTML BODY = 'a:\quantilesoutput-
  body.htm'
  CONTENTS = 'a:\quantilesoutput-
  contents.htm'
  PAGE = 'a:\quantilesoutput-page.htm'
  FRAME = 'a:\quantilesoutput-frame.htm';
```
The ODS HTML statement sets up the files to receive the output to the HTML destination.

```
PROC PRINT DATA = abcxyz.quantilesoutput;
TITLE 'quantilesoutput dataset';
```

```
RUN;
```
The RUN statement at this point in the program forces SAS to carry out the PROC PRINT before considering the rest of the program.

```
ODS HTML CLOSE;
```
No more output to the HTML destination is required, so the destination is closed.

```
ODS LISTING;
```
The statement ODS LISTING restores the default open status of this destination.

```
RUN;
```

SAS Output for Program 8.1.3

Quantiles Output Dataset

Obs	VarName	Quantile	Estimate
1	rep1	100% Max	10150.00
2	rep1	99%	10150.00
3	rep1	95%	6637.00
4	rep1	90%	3801.00
5	rep1	75% Q3	935.40
6	rep1	50% Median	238.00
7	rep1	25% Q1	−0.01
8	rep1	10%	−9.41
9	rep1	5%	−10.94
10	rep1	1%	−14.40
11	rep1	0% Min	−14.40
12	tnfaver	100% Max	9601.00
13	tnfaver	99%	9601.00
14	tnfaver	95%	6058.67
15	tnfaver	90%	3531.33
16	tnfaver	75% Q3	988.50
17	tnfaver	50% Median	209.85
18	tnfaver	25% Q1	8.30
19	tnfaver	10%	−4.72
20	tnfaver	5%	−7.34
21	tnfaver	1%	−13.80
22	tnfaver	0% Min	−13.80

8.2   SAVING SAS OUTPUT AS A SAS DATASET

Many SAS procedures enable the user to save the numeric parts of the output produced as a SAS dataset. In Program 8.2.1, PROC FREQ is used to create a dataset "outtab" consisting of the counts and percentages that would appear in the cross-tabulation normally output. As the object of running the procedure is to obtain the output dataset, the NOPRINT option is used to suppress the printing of the table.

This use of OUT = to output a dataset from a procedure is a feature of SAS that can prove to be useful in some analyses where further manipulations of the output dataset are undertaken.

PROGRAM 8.2.1

```
Comments omitted
LIBNAME abcxyz 'a:\data';
LIBNAME library 'a:\formats';

PROC FREQ DATA = abcxyz.tnfdata;
TABLES mtb*fat/NOPRINT OUT = outtab;
The NOPRINT option suppresses any output from this pro-
cedure. The numbers that would be used to create the PROC
FREQ table are output to a new dataset called "outtab" as
a result of the OUT = OUTTAB option.

PROC PRINT DATA = outtab;
The "outtab" dataset is printed.

RUN;
```

SAS Output for Program 8.2.1

Obs	mtb	fat	COUNT	PERCENT
1	No MTB	No FAT	11	25
2	No MTB	FAT present	11	25
3	MTB present	No FAT	11	25
4	MTB present	FAT present	11	25

8.3   TASKS

TASK 8.1

Carry out some of the tasks from Chapter 2 to Chapter 7 using the ODS destinations HTML, RTF, and PDF.

TASK 8.2

Repeat Task 6.4, with the addition of the OUT = option to save the PROC FREQ table as a dataset. Use PROC PRINT to view this dataset.

9 Graphics in SAS

SAS has the ability to produce many different sorts of graphs and charts. Many of these are available through the SAS/GRAPH module, but several other procedures also have the ability to produce high-resolution graphics. This section contains some examples of what SAS can produce. SAS provides the user with several ways of customising the graphs produced, and what follows is just a sample of the possibilities.

9.1 LOS ANGELES DEPRESSION DATA

Data on 49 respondents in a study on depression are shown in Appendix B. The code book for the data and a SAS program that will read the plain text data and save it as a permanent dataset are also shown.

9.2 HISTOGRAM OF CESD SCORE

In the data are self-attributed scores on 20 statements designed to indicate an individual's level of depression. The SAS program in Appendix B gives the text of the statements. The responses to the statements are scored 0 for "rarely or none of the time (less than 1 day)"; 1 for "some or a little of the time (1–2 days)"; 2 for "occasionally or a moderate amount of the time (3–4 days)"; 3 for "most or all of the time (5–7 days)". The sum of these scores over the 20 statements gives the Center for Epidemiologic Studies Depression Scale (CESD) score.

9.2.1 BASIC HISTOGRAM

Program 9.2.1 shows the code necessary to create a basic histogram with PROC UNIVARIATE.

PROGRAM **9.2.1**

```
Comments omitted
LIBNAME abcxyz 'a:\data';
LIBNAME library 'a:\formats';

PROC UNIVARIATE DATA=abcxyz.depress
   NOPRINT;
VAR cesd;
HISTOGRAM cesd;
```
The NOPRINT option suppresses any of the normal output from this procedure. The VAR statement tells PROC UNIVARIATE which variable to summarise and the HISTOGRAM statement creates the plot.
```
RUN;
```

SAS Output for Program 9.2.1

　See Figure 9.1.

9.2.2   LABELLING OF END-POINTS FOR BARS

Histograms are more usually shown with the end-points of the bars labelled, rather than the mid-points. Program 9.2.2. shows how a histogram of this sort can be created in SAS.

PROGRAM **9.2.2**

```
Comments omitted
LIBNAME abcxyz 'a:\data';
LIBNAME library 'a:\formats';
```

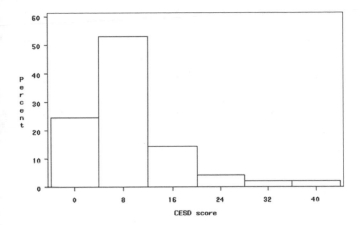

FIGURE 9.1 Basic histogram of CESD score.

```
PROC UNIVARIATE DATA=abcxyz.depress
   NOPRINT;
VAR cesd;
HISTOGRAM cesd/ENDPOINTS;
```
The ENDPOINTS option causes SAS to create bars according to end-points rather than the mid-points.
```
RUN;
```

SAS Output for Program 9.2.2

See Figure 9.2.

9.2.3 DEFINING THE BARS

The histograms shown in Figure 9.1 and Figure 9.2 have had their choice of bars decided by SAS. Program 9.2.3 shows how the user can control this.

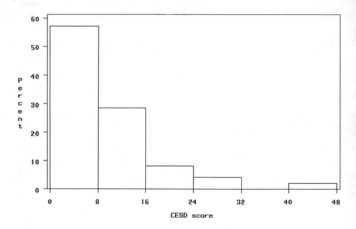

FIGURE 9.2 Histogram of CESD score with end-points for bars.

PROGRAM 9.2.3

```
Comments omitted
LIBNAME abcxyz 'a:\data';
LIBNAME library 'a:\formats';
PROC UNIVARIATE DATA=abcxyz.depress
  NOPRINT;
VAR cesd;
HISTOGRAM cesd/ENDPOINTS=0 TO 45 BY 5;
An extension of the ENDPOINTS option causes SAS to use bars
starting at 0, 5, 10, 15, …, 40 and finishing with 45.
RUN;
```

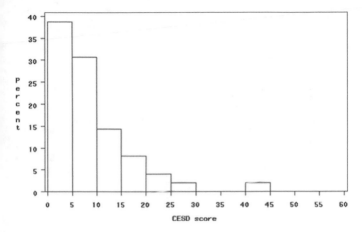

FIGURE 9.3 Histogram of CESD score with bars defined.

SAS Output for Program 9.2.3

See Figure 9.3.

9.2.4 COUNTS AND LABELS ON THE VERTICAL AXIS

The vertical axes of the histograms in Figure 9.1, to Figure 9.3 are in terms of percentages. In Program 9.2.4, this is changed to counts and a label is applied.

PROGRAM 9.2.4

```
Comments omitted
  LIBNAME abcxyz 'a:\data';
  LIBNAME library 'a:\formats';
  PROC UNIVARIATE DATA=abcxyz.depress
    NOPRINT;
```

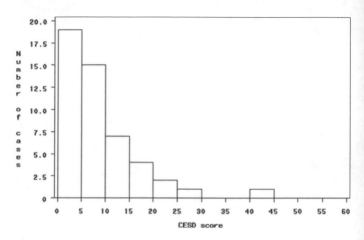

FIGURE 9.4 Histogram of CESD score with counts and labels.

```
VAR cesd;
HISTOGRAM cesd/ENDPOINTS=0 TO 45 BY 5
  VSCALE=COUNT
VAXISLABEL='Number of cases';
```
The VSCALE option tells SAS to use counts for the vertical axis and the VAXISLABEL option gives a label to the vertical axis.
```
RUN;
```

SAS Output for Program 9.2.4

See Figure 9.4.

9.2.5 ADDING A NORMAL CURVE

Frequently the purpose of a histogram is to compare the distribution of the graphed variable to the Normal distribution. Program 9.2.5 makes SAS add a Normal curve to the histogram to facilitate this comparison.

PROGRAM 9.2.5

```
Comments omitted
LIBNAME abcxyz 'a:\data';
LIBNAME library 'a:\formats';
PROC UNIVARIATE DATA=abcxyz.depress
   NOPRINT;
VAR cesd;
HISTOGRAM cesd/ENDPOINTS=0 TO 45 BY 5
   VSCALE=COUNT VAXISLABEL='Number of
   cases'
   NORMAL(NOPRINT COLOR=BLACK);
The NORMAL option asks for the Normal curve to be added.
The (NOPRINT COLOR=BLACK) additional options sup-
press the display of various statistics that can also be used
to compare the distribution of the variable with the Normal
distribution and ask for the curve to be black.
RUN;
```

SAS Output for Program 9.2.5

See Figure 9.5.

9.3 BAR CHART OF MARITAL STATUS

In the Los Angeles Depression Data, one of the variables recorded is
marital status. This is coded 1 for "never married"; 2 for "married", 3
for "divorced", 4 for "separated", 5 for "widowed".

9.3.1 BASIC BAR CHART

Program 9.3.1 shows the code necessary to create a basic vertical bar
chart with PROC GCHART.

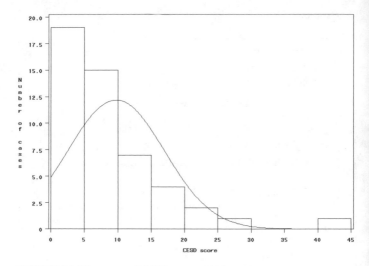

FIGURE 9.5 Histogram of CESD score with Normal curve.

PROGRAM 9.3.1

```
Comments omitted

LIBNAME  abcxyz  'a:\data';
LIBNAME  library  'a:\formats';

PROC  GCHART  DATA=abcxyz.depress;
VBAR  marital/DISCRETE;
PATTERN  COLOR=BLACK  VALUE=EMPTY;
```
The VBAR statement requests a vertical bar chart of the "marital" variable. The DISCRETE option tells SAS that "marital" is a discrete variable. If this is omitted, SAS will operate as if the variable is continuous and will create categories of its own for the bars.

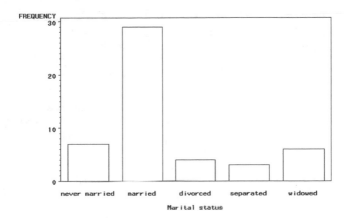

FIGURE 9.6 Basic bar chart of marital status.

> *The PATTERN statement defines the colour of the bars. The VALUE option here defines the bars to be empty.*
>
> RUN;

SAS Output for Program 9.3.1

See Figure 9.6.

9.3.2 THREE-DIMENSIONAL BAR CHART

Program 9.3.2 shows how the bar chart in Figure 9.6 can be represented in a three-dimensional style.

PROGRAM 9.3.2

```
Comments omitted
LIBNAME abcxyz 'a:\data';
LIBNAME library 'a:\formats';
```

```
PROC GCHART DATA=abcxyz.depress;
VBAR3D marital/DISCRETE SHAPE=CYLINDER;
PATTERN COLOR=GRAYCC VALUE=SOLID;
```
*The VBAR3D statement requests a three-dimensional vertical
bar chart of the "marital" variable. The SHAPE option requests
the bars to be cylinders.*

*The PATTERN statement defines gray bars with darkness con-
trolled by the hexadecimal number "CC" where "FF" would
give white and "00" would give black. Colours can be controlled
by using names (e.g. "RED") and "CXrrggbb" (where "rr", "gg"
and "bb" are hexadecimal numbers controlling the amount of
red, green and blue). The VALUE option here defines the bars
to be solid.*
```
RUN;
```

SAS Output for Program 9.3.2

See Figure 9.7.

9.3.3 BAR CHART WITH FORMATTED AXIS

In Program 9.3.3, the two-dimensional bar chart of Figure 9.6 is repeated
with a formatted vertical axis and a different pattern for the bars.

PROGRAM 9.3.3

```
Comments omitted

LIBNAME abcxyz 'a:\data';
LIBNAME library 'a:\formats';
PROC GCHART DATA=abcxyz.depress;
AXIS1 MAJOR=(NUMBER=7) LABEL=('Count');
VBAR marital/DISCRETE AXIS=AXIS1;
```

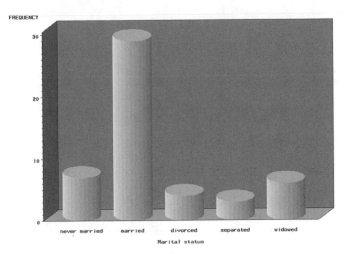

FIGURE 9.7 Three-dimensional bar chart of marital status.

```
PATTERN COLOR=GRAY5C VALUE=L1;
```

The AXIS1 statement prepares a format for an axis with 7 major tick marks (from the MAJOR=(NUMBER=7)) and a label. Further axis statements can be used for other graphs using the statements AXIS2, AXIS3, …, to define the formats. In the VBAR statement, the AXIS=AXIS1 option defines the format of the vertical axis to be that defined by the AXIS1 statement.

The PATTERN statement defines the colour and the VALUE option here defines the bars to have left slanting hatching with intensity 1. The intensity can vary from 1 to 5.

```
RUN;
```

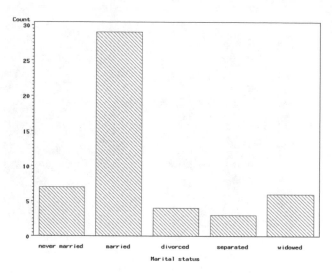

FIGURE 9.8 Bar chart of marital status with formatted axis.

SAS Output for Program 9.3.3

See Figure 9.8.

9.3.4   BAR CHART WITH FORMATTED AXIS AND LABELLED BARS

In Program 9.3.4, labels are added to the bar chart of Figure 9.8.

PROGRAM 9.3.4

```
Comments omitted
LIBNAME abcxyz 'a:\data';
LIBNAME library 'a:\formats';

PROC GCHART DATA=abcxyz.depress;
AXIS1 MAJOR=(NUMBER=7) LABEL=('Count');
```

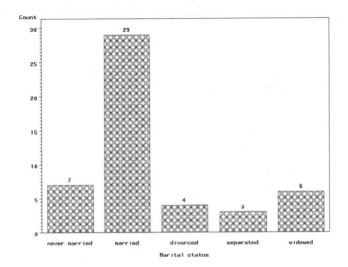

FIGURE 9.9 Bar chart of marital status with formatted axis and labeled bars.

```
VBAR marital/DISCRETE AXIS=AXIS1
   OUTSIDE=FREQ;

PATTERN COLOR=GRAY5C VALUE=X4;
```
*The OUTSIDE=FREQ option in the VBAR statement
requests labels for the bars to be the frequency count, positioned outside the bars.*
*The PATTERN statement defines the colour and the VALUE
option here defines the bars to have cross-hatching with
intensity 4.*
```
RUN;
```

SAS Output for Program 9.3.4

See Figure 9.9.

9.3.5 Bar Chart with Summaries by Another Variable

Program 9.3.5 produces a bar chart where the height of the bars correspond to values of the CESD score.

Program 9.3.5

```
Comments omitted
LIBNAME abcxyz 'a:\data';
LIBNAME library 'a:\formats';

PROC GCHART DATA=abcxyz.depress;
VBAR marital/DISCRETE SUMVAR=cesd
   TYPE=MEAN OUTSIDE=MEAN;
PATTERN COLOR=BLACK VALUE=R1;
```
The SUMVAR option for the VBAR statement defines the variable that is summarised. The TYPE=MEAN option defines the summary statistic to be the mean and the OUTSIDE=MEAN option labels the bars with the mean of the summary variable, outside the bars
The PATTERN statement defines another colour and the VALUE option here defines the bars to have right-slanting hatching with intensity 1.
```
RUN;
```

SAS Output for Program 9.3.5

See Figure 9.10.

9.4 BAR CHART OF RELIGION GROUPED BY MARITAL STATUS

A person's religious affiliation was also recorded in the Los Angeles Depression Data. This is coded 1 for "Protestant", 2 for "Catholic", 3

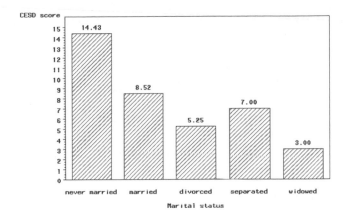

FIGURE 9.10 Bar chart of marital status with summaries by variable.

tor "Jewish", 4 for "none", 5 for "other". In the 49 cases in Appendix B, there are no instances of "Jewish" or "other".

9.4.1 HORIZONTAL BAR CHART

A horizontal bar chart has been used in Program 9.4.1 to show bar charts of religion, grouped by marital status because it allows for summary statistics to be positioned conveniently.

PROGRAM 9.4.1

Comments omitted

```
LIBNAME abcxyz 'a:\data';
LIBNAME library 'a:\formats';
PROC GCHART DATA=abcxyz.depress;
HBARreligion/DISCRETE GROUP=marital;
```
The HBAR statement requests a horizontal bar chart of the

> *"religion" variable. The GROUP option specifies the*
> *grouping variable.*
>
> RUN;

SAS Output for Program 9.4.1

See Figure 9.11.

9.4.2 STACKED BAR CHART

A bar chart of religion is produced by Program 9.4.2, showing marital status stacked within each of the bars.

PROGRAM 9.4.2

```
Comments omitted
LIBNAME  abcxyz  'a:\data';
LIBNAME  library  'a:\formats';

PROC GCHART DATA=abcxyz.depress;
VBAR religion/DISCRETE SUBGROUP=marital;
PATTERN1  COLOR=GRAY5C  VALUE=R1;
PATTERN2  COLOR=GRAY5C  VALUE=R2;
PATTERN3  COLOR=GRAY5C  VALUE=R3;
PATTERN4  COLOR=GRAY5C  VALUE=L3;
PATTERN5  COLOR=GRAY5C  VALUE=X3;
```
The VBAR statement requests a vertical bar chart of the
"religion" variable. The SUBGROUP option specifies the
grouping variable.
The PATTERN1, PATTERN2, ..., PATTERN5 statements
define the colour and fill characteristics for each category
of the subgroup variable.
```
RUN;
```

SAS Output for Program 9.4.2

See Figure 9.12.

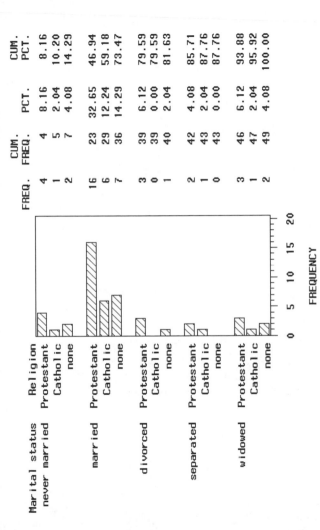

Marital status	Religion	FREQ.	CUM. FREQ.	PCT.	CUM. PCT.
never married	Protestant	4	4	8.16	8.16
	Catholic	1	5	2.04	10.20
	none	2	7	4.08	14.29
married	Protestant	16	23	32.65	46.94
	Catholic	6	29	12.24	59.18
	none	7	36	14.29	73.47
divorced	Protestant	3	39	6.12	79.59
	Catholic	0	39	0.00	79.59
	none	1	40	2.04	81.63
separated	Protestant	2	42	4.08	85.71
	Catholic	1	43	2.04	87.76
	none	0	43	0.00	87.76
widowed	Protestant	3	46	6.12	93.88
	Catholic	1	47	2.04	95.92
	none	2	49	4.08	100.00

FIGURE 9.11 Horizontal bar chart of religion grouped by marital status.

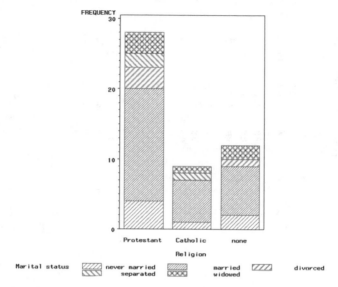

FIGURE 9.12 Stacked bar chart of religion grouped by marital status.

9.5 PIE CHART OF MARITAL STATUS

9.5.1 BASIC PIE CHART

Program 9.5.1 shows the code necessary to create a basic pie chart with PROC GCHART.

PROGRAM **9.5.1**

```
Comments omitted
LIBNAME abcxyz 'a:\data';
LIBNAME library 'a:\formats';PROC GCHART
DATA=abcxyz.depress;
PIE marital/DISCRETE
PATTERN1  COLOR=GRAYC0  VALUE=PSOLID;
PATTERN2  COLOR=GRAY90  VALUE=PSOLID;
PATTERN3  COLOR=GRAT60  VALUE=PSOLID;
PATTERN4  COLOR=GRAY30  VALUE=PSOLID;
PATTERN5  COLOR=GRAY00  VALUE=PSOLID;
```
The PIE statement requests a pie chart of the "marital"
variable. The DISCRETE option tells SAS that "marital" is
a discrete variable. If this is omitted, SAS will operate as if
the variable is continuous and will create categories of its
own for the pie slices.
The PATTERN1, PATTERN2, ..., PATTERN5 statements
define the colour and fill characteristics for each slice of the
pie.
```
RUN;
```

SAS Output for Program 9.5.1

See Figure 9.13.

9.5.2   PIE CHART WITH FORMATTING

Some formatting can be applied to Figure 9.13. The code is shown in
Program 9.5.2.

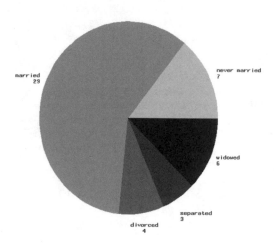

FIGURE 9.13 Pie chart of marital status.

PROGRAM 9.5.2

```
Comments omitted
LIBNAME abcxyz 'a:\data';
LIBNAME library 'a:\formats';

PROC GCHART DATA=abcxyz.depress;
PIE marital/DISCRETE VALUE=ARROW ANGLE
  =90 TYPE=PERCENT;
PATTERN1 COLOR=GRAYC0 VALUE=PSOLID;
PATTERN2 COLOR=GRAY90 VALUE=PSOLID;
PATTERN3 COLOR=GRAT60 VALUE=PSOLID;
PATTERN4 COLOR=GRAY30 VALUE=PSOLID;
PATTERN5 COLOR=GRAY00 VALUE=PSOLID;
```

> *The VALUE=ARROW option in the PIE statement tells SAS to provide an arrow from the pie slice descriptions to the pie slices themselves. The ANGLE=90 option is used to tell SAS at what angle the first pie slice should be positioned. Without it here, some of the pie slice descriptions partially obscure each other. The TYPE=PERCENT option requests that percentages are shown in the descriptions instead of counts.*
>
> RUN;

SAS Output for Program 9.5.2

See Figure 9.14.

9.5.3 PIE CHART WITH FORMATTING AND AN EXPLODED SEGMENT

An "exploded segment" in a pie chart is a segment that has been partially moved away from the centre of the pie. Program 9.5.3 creates a chart containing this feature.

PROGRAM 9.5.3

```
Comments omitted
LIBNAME  abcxyz  'a:\data';
LIBNAME  library  'a:\formats';
PROC GCHART DATA=abcxyz.depress;
PIE marital/DISCRETE  VALUE=ARROW
   ANGLE = 90 TYPE=PERCENT EXPLODE = 1;
PATTERN1  COLOR=GRAYC0  VALUE=PSOLID;
PATTERN2  COLOR=GRAY90  VALUE=PSOLID;
PATTERN3  COLOR=GRAT60  VALUE=PSOLID;
PATTERN4  COLOR=GRAY30  VALUE=PSOLID;
PATTERN5  COLOR=GRAY00  VALUE=PSOLID;
```

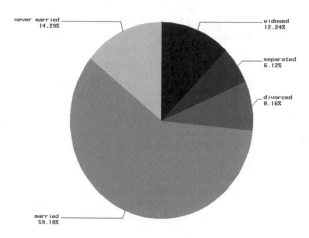

FIGURE 9.14 Formatted pie chart of marital status.

> *The EXPLODE=1 option tells SAS to "explode" the segment associated with the first category of the variable "marital".*
>
> RUN;

SAS Output for Program 9.5.3

See Figure 9.15.

9.6 PLOT OF CESD SCORE AGAINST AGE

The ages of the respondents were also recorded in the Los Angeles Depression Data in Appendix B. A plot of CESD Score against age may help to identify any relationship between the two variables.

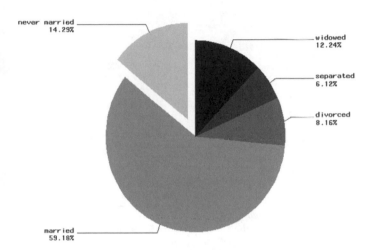

FIGURE 9.15 Pie chart of marital status with exploded segment.

9.6.1 BASIC SCATTER PLOT

A scatter plot of the two variables can be obtained using PROC GPLOT, as in Program 9.6.1.

PROGRAM 9.6.1

```
Comments omitted
LIBNAME abcxyz 'a:\data';
LIBNAME library 'a:\formats';

PROC GPLOT DATA=abcxyz.depress;
PLOT cesd*age;
```

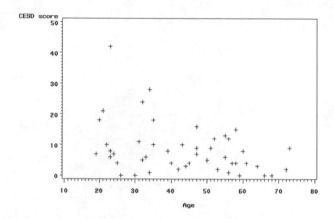

FIGURE 9.16 Plot of CESD score against age.

> *The PLOT statement defines the variables to be plotted. The*
> *variable before the "*" is put on the vertical axis and the*
> *variable following it is put on the horizontal axis.*
>
> RUN;

● *SAS Output for Program 9.6.1*

See Figure 9.16.

9.6.2   SCATTER PLOT WITH FORMATTING

The scatter plot in Figure 9.16 can be enhanced using some formatting, as in Program 9.6.2.

PROGRAM 9.6.2

> *Comments omitted*
>
> LIBNAME abcxyz 'a:\data';

```
LIBNAME library 'a:\formats';
PROC GPOT DATA=abcxyz2.depress;
SYMBOL1 VALUE=STAR
PLOT cesd*age=1/HAXIS=15 TO 75 BY 5
  HMINOR=0 VAXIS=0 TO 45 BY 5
  VMINOR=0;
```
The SYMBOL1 statement prepares a format for a symbol with the VALUE=STAR option defining the symbol itself to be a star. Further symbol statements can be used for other graphs using the statements SYMBOL2, SYMBOL3, ..., to define the formats.

The "cesd*age=1" part of the PLOT statement tells SAS to use the symbol format defined by the SYMBOL1 when creating the plot. The HAXIS and VAXIS options define the range of the horizontal and vertical axes respectively and the position of the major tick marks. The HMINOR=0 and VMINOR=0 tell SAS to include no minor tick marks between the major tick marks.

```
RUN;
```

SAS Output for Program 9.6.2

See Figure 9.17.

9.7 TASKS

These tasks refer to the Los Angeles Depression Data in Appendix B.

TASK 9.1

Create a histogram for the "income" variable. Apply formatting to the vertical axis, define the bar end-points and superimpose a Normal curve."

TASK 9.2

Create a horizontal bar chart for the "education" variable, summarised by the "income" variable.

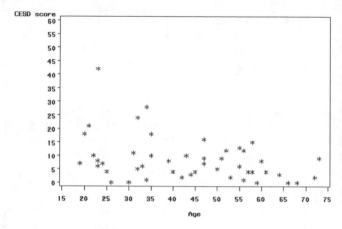

FIGURE 9.17 Scatter plot with formatting.

TASK 9.3

Create a horizontal stacked bar chart for the "education" variable, grouped by the "employment" variable.

TASK 9.4

Create a pie chart for the "education" variable and apply formatting.

10 An Introduction to Macros

Macros are a group of statements that can be executed several times with just a few bits of code (often called "arguments") different each time. The macro statements are often kept in a file that is separate from the program that actually invokes the macro.

The examples shown in Section 10.1 and Section 10.2 are relatively simple examples of the SAS macro language, but the opportunities offered are great..

10.1 EXAMPLE OF A MACRO TO PRINT DATASETS

The TNF data have previously been seen in Table 4.2 as four separate datasets: one for MTB and FAT both not present; one for MTB present, FAT not present; one for MTB not present, FAT present; one for MTB and FAT both present. These four datasets have been stored as permanent datasets "tnfdata1", "tnfdata2", "tnfdata3" and "tnfdata4" by Programs 4.1.1, 4.1.2, 4.1.3 and 4.1.4.

The SAS code to print each of these datasets with appropriate titles is given in Program 10.1.1.

PROGRAM 10.1.1

```
Comments omitted
LIBNAME abcxyz 'a:\data';
PROC PRINT DATA=abcxyz.tnfdata1;
TITLE 'Data for dataset with MTB not
  present and FAT not present';
VAR donor rep1 rep2 rep3;
```

```
PROC PRINT DATA=abcxyz.tnfdata2;
TITLE 'Data for dataset with MTB present
   and FAT not present';
VAR donor rep1 rep2 rep3;
PROC PRINT DATA=abcxyz.tnfdata3;
TITLE 'Data for dataset with MTB not
   present and FAT present';
VAR donor rep1 rep2 rep3;
PROC PRINT DATA=abcxyz.tnfdata4;
TITLE 'Data for dataset with MTB present
   and FAT present';
VAR donor rep1 rep2 rep3;

RUN;
```

The repetitious code seen in Program 10.1.1 is slightly tedious for even this very simple procedure and small number of datasets. In situations where there are perhaps more datasets and the procedure to be repeated is lengthier, the repetition of the code is also rather inefficient.

An alternative and more efficient way of proceeding would be to use macro 10.1.1 which could be contained in a file "a:\macro10_1_1.sas".

Macro 10.1.1

```
Comments omitted
LIBNAME abcxyz 'a:\data';
%MACRO datprint(dataset,title);
PROC PRINT DATA=&dataset;
TITLE "Data for dataset with &title";
```

> *Note the use of double quotation marks. If single quotation marks are used, the text "&title" appears in the title of the PROC PRINT rather than the required replacement text.*
>
> ```
> VAR donor rep1 rep2 rep3;
>
> %MEND;
> ```

The "dataset" and "title" are arguments that are passed to the macro by Program 10.1.2.

PROGRAM 10.1.2

> *Comments omitted*
>
> ```
> OPTIONS MPRINT SYMBOLGEN;
> ```
> *These options make debugging of the macros easier. The MPRINT option shows the steps of the macro in the log and the SYMBOLGEN option displays in the log how the arguments are handled.*
>
> ```
> %INCLUDE 'a:\macro10_1_1.sas';
> ```
> *The INCLUDE statement tells SAS to open up the macro, ready for use.*
>
> ```
> %datprint(abcxyz.tnfdata1,MTB not
> present and FAT not present)
> %datprint(abcxyz.tnfdata2,MTB present
> and FAT not present)
> %datprint(abcxyz.tnfdata3,MTB not
> present and FAT present)
> %datprint(abcxyz.tnfdata4,MTB present
> and FAT present)
>
> RUN;
> ```

Additional Notes for Program 10.1.2

The "%datprint" statements tells SAS to carry out the commands contained in the "datprint" macro. The %INCLUDE statement has allowed SAS to have access to this macro. Note that it is not necessary to have a semi-colon to terminate a macro call.

The first "%datprint" statement passes two arguments to the macro: "abcxyz.tnfdata1" and "MTB not present and FAT not present". In the macro, the first argument is used wherever "&dataset" appears in the macro and the second argument is used wherever "&title" appears in the macro. Thus the code.

```
PROC PRINT DATA=&dataset;
TITLE "Data for dataset with &title";
VAR donor rep1 rep2 rep3;
```

effectively becomes

```
PROC PRINT DATA=abcxyz.tnfdata1;
TITLE "Data for dataset with MTB not present
   and FAT not present";
VAR donor rep1 rep2 rep3;
```

10.2 EXAMPLE OF A MACRO TO CREATE AND FORMAT A HISTOGRAM

Program 9.2.5 produces a histogram for the CESD score from the Los Angeles Depression Data of Appendix B, formatting it with defined endpoints for the bars, requesting counts to be shown, a title specified for the vertical axis and a Normal curve superimposed.

Macro 10.2.1 (contained in file "a:\macro10_2_1.sas") and Program 10.2.1 can be used to create the same histogram as in Figure 9.5 and also simply produce other histograms.

MACRO 10.2.1

```
Comments omitted
LIBNAME abcxyz 'a:\data';
%MACRO hist(dataset,variable,from,to,by,
  vscale,vlabel,normal,title);
PROC UNIVARIATE DATA=&dataset NOPRINT;
TITLE &title;
VAR &variable;
HISTOGRAM &variable/ENDPOINTS=&from TO
  &to BY & by VSCALE=&vscale
  VAXISLABEL=&vlabel&normal;
%MEND;
```

There are nine arguments that are passed to the macro by Program 10.2.1: "dataset", "variable", "from", "to", "by", "vscale", "vlabel", "normal", "title".

PROGRAM 10.2.1

```
Comments omitted
OPTIONS MPRINT SYMBOLGEN;
%INCLUDE 'a:\macro10_2_1.sas';
%hist(abcxyz.depress,cesd,0,45,5,count,
   'Number of cases',normal(noprint
   color=black),
   'Histogram of CESD score')
```
Passing these arguments to the macro will produce the histogram shown in Figure 10.1 which is the same as figure 9.5 produced by Program 9.2.5, with the addition of a title.
```
%hist (abcxyz.depress, cesd, 0,45,5, count,
   'Number of cases',,
   'Histogram of CESD score')
```

> *The existence of nothing for the eighth argument (the seventh being "Number of cases" and the ninth being "Histogram of CESD score") means that the macro does not use the code that requests the Normal curve on the histogram. The resulting histogram is shown in Figure 10.2.*
>
> ```
> %hist(abcxyz.depress,age,15,75,10,
>     [percent, 'Percentage',,]
>
>     'Histogram of Age')
> ```
>
> *A histogram for the "age" variable is requested with the scale of the vertical axis specified as 15 to 70 with major tick marks every 10 units. Percentages are requested for the vertical axis with an appropriate label and a new title. The resulting histogram is shown in Figure 10.3.*
>
> RUN;

SAS Output for Program 10.2.1

See Figure 10.1, Figure 10.2, and Figure 10.3.

10.3 TASKS

Task 10.1

For the Nurses' Glove Use Data in Appendix A, the file "glove-sonerow.dat" contains the data with one row per nurse. The first variable in this file is the nurse's number; the next four variables are the number of times the nurse was observed before the educational programme and at 1, 2 and 5 months after the programme; the next four variables are the number of times that the nurse used gloves in each of these observation periods; the last variable is the number of years experience that the nurse has.

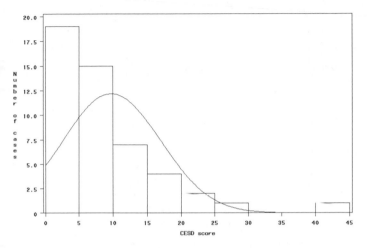

Figure 10.1 Histogram of CESD score with Normal curve from macro.

Apart from the nurse's number and the number of years experience, the remaining eight variables can be seen as occurring in four pairs, namely, the number of observation sessions and number of times that gloves are used: (i) before the programme; (ii) 1 month after the programme; (iii) 2 months after the programme; (iv) 5 months after the programme.

Construct a macro that produces a crosstabulation (using PROC FREQ) of whatever pair of variables are requested by a separate program.

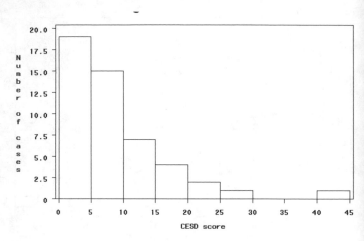

FIGURE 10.2 Histogram of CESD score without Normal curve from macro.

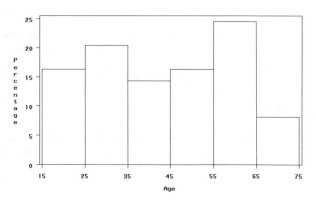

FIGURE 10.3 Histogram of age from macro.

TASK 10.2

The Los Angeles Depression Data in Appendix B contains data on 20 items from a depression scale (variables "statement1" to "statement20"). Produce a macro that produces a bar chart for whichever of these statements is requested by a separate program.

Appendix A

Nurses' Glove Use Data

An experiment was undertaken to assess how an educational programme affected the use of gloves by nurses in an emergency paediatric department. Without their knowledge, nurses were observed during "vascular access procedures" before the educational programme, and again 1, 2, and 5 months after the programme. How often the nurses wore gloves was recorded. Experience is in years.

TABLE A.1

Nurse	Period	Observed	Gloves	Experience
1	1	2	1	15
1	2	7	6	15
1	3	1	1	15
1	4			15
2	1	2	1	2
2	2	6	5	2
2	3	11	10	2
2	4	9	9	2
3	1	5	5	3
3	2	13	13	3
3	3	8	7	3
3	4	15	14	3
4	1	2	0	10
4	2	2	2	10
4	3	2	2	10
4	4	5	4	10
5	1	12	0	20
5	2	2	2	20
5	3	3	3	20
5	4	3	0	20
6	1	3	0	8
6	2	8	8	8
6	3	3	2	8
6	4	4	2	8
7	1	4	4	9
7	2	4	4	9
7	3			9
7	4			9
8	1	4	0	9
8	2	4	4	9
8	3	2	1	9
8	4			9
9	1	2	0	15

TABLE A.1 (continued)

Nurse	Period	Observed	Gloves	Experience
9	2	3	2	15
9	3	1	1	15
9	4	2	1	15
10	1	6	1	8
10	2	1	1	8
10	3	2	2	8
10	4			8
11	1	3	0	8
11	2	4	3	8
11	3	8	6	8
11	4	2	0	8
12	1	2	0	2
12	2	3	3	2
12	3	8	8	2
12	4	5	5	2
13	1	1	0	5
13	2			5
13	3			5
13	4			5
14	1	1	0	15
14	2	3	3	15
14	3			15
14	4			15
15	1	1	1	3
15	2	2	2	3
15	3	1	1	3
15	4	1	1	3
16	1	1	0	14
16	2			14
16	3			14
16	4	1	1	14
17	1			14
17	2	2	2	14

TABLE A.1 (continued)

Nurse	Period	Observed	Gloves	Experience
17	3	3	3	14
17	4	1	1	14
18	1			8
18	2	1	1	8
18	3	1	1	8
18	4	1	1	8
19	1			3
19	2	2	1	3
19	3			3
19	4			3
20	1			6
20	2	1	1	6
20	3			6
20	4			6
21	1			3
21	2	1	1	3
21	3			3
21	4			3
22	1			1
22	2			1
22	3	2	2	1
22	4			1
23	1			6
23	2			6
23	3			6
23	4	1	0	6

Source: Data were used in Friedland, L., Joffe, M., Moore, D. et al., *Am. J. Dis. Child.*, 146, 1355–1358, 1992. Obtained from *The Data and Story Library*, http://lib.stat.cmu.edu/DASL/, and used here with kind permission of the authors of the article.

Appendix B

Los Angeles Depression Data

A study on depression was undertaken, and data from a survey given to 49 respondents are shown in Table B.1. The meanings of the variables are listed in Table B.2.

A SAS program to read in the raw data is given in Program B.1.

TABLE B.1

```
 1  1 2 68 5 2 4 4 1 0 0 0 0 0 0 0 0 0 0 0 0 0 0 0 0 0 0 0 0 0 0 0 0 2 2 1 1 0 0 1
 2  2 1 58 3 4 1 15 1 0 0 1 0 0 0 0 0 0 0 1 0 0 1 0 1 0 0 0 4 0 1 1 1 1 0 0 1
 3  3 2 45 2 3 1 28 1 0 0 0 0 1 0 0 0 0 0 0 0 1 1 1 0 0 0 0 4 0 1 2 1 1 0 0 0
 4  4 2 50 3 3 3 39 1 0 0 0 0 1 1 0 3 0 0 0 0 0 0 0 0 0 0 0 5 0 2 1 1 2 0 0 1
 5  5 2 33 4 3 1 35 1 0 0 0 0 0 0 3 3 0 0 0 0 0 0 0 0 0 0 0 6 0 1 1 1 1 1 1 0
 6  6 1 24 2 3 1 11 1 0 0 0 0 0 0 0 1 0 0 1 2 0 0 2 1 0 0 0 7 0 1 1 1 1 0 1 1
 7  7 2 58 2 2 5 11 1 2 1 1 2 1 0 0 2 2 0 0 0 0 0 3 0 0 0 0 1 15 0 2 3 1 1 0 1 1
 8  8 1 22 1 3 191 0 1 2 0 2 1 0 0 0 0 0 0 0 0 1 1 1 1 0 10 0 2 1 2 2 0 1 0
 9  9 2 47 2 3 4 23 2 0 1 0 0 3 0 0 0 0 0 3 0 3 2 3 0 0 0 0 16 1 1 4 1 1 1 0 1
10 10 1 30 2 2 1 35 4 0 0 0 0 0 0 0 0 0 0 0 0 0 0 0 0 0 0 0 1 1 1 2 0 0 0
11 11 2 20 1 2 3 25 4 0 0 1 0 1 2 1 0 0 1 0 1 2 2 1 1 2 3 0 0 18 1 1 2 1 2 0 0 0
12 12 2 57 2 3 2 24 1 0 0 0 0 0 0 0 0 0 0 0 0 0 0 2 2 0 0 0 0 4 0 2 2 1 1 1 1 1
13 13 1 39 2 2 1 28 1 1 1 0 0 0 0 0 0 0 0 0 1 0 2 0 1 0 0 1 1 8 0 1 3 1 1 0 1 0
14 14 2 61 5 3 4 13 1 0 0 0 0 1 0 0 1 0 0 0 1 0 0 0 0 0 0 0 1 4 0 1 1 1 1 0 1 0
15 15 2 23 2 3 1 15 2 0 0 0 0 0 0 0 0 0 0 0 0 0 1 3 1 0 2 1 8 0 1 1 1 2 0 0 0
16 16 2 21 1 2 1 6 1 1 1 2 0 1 1 1 1 2 2 0 1 1 2 1 1 1 2 0 0 21 1 1 3 1 1 1 0 1
17 17 2 23 1 4 1 8 1 3 3 2 3 3 3 2 2 3 2 2 2 1 2 3 2 0 1 0 3 42 1 1 1 2 2 1 1 0
18 18 2 55 4 2 3 191 1 0 1 1 1 0 0 0 0 0 2 0 0 0 0 0 0 0 0 0 6 0 2 3 1 1 1 1 1
```

```
19 19 2 26 1 6 1 15 1 0 0 0 0 0 0 0 0 0 0 0 0 0 0 0 0 0 0 0 0 0 0 2 2 2 2 1 1 0
20 20 1 64 5 2 494 0 0 0 0 0 0 3 0 0 0 0 0 0 0 0 0 0 0 0 3 0 1 2 1 2 0 0 0
21 21 2 44 1 3 1 6 2 0 0 0 0 0 0 0 0 3 0 0 0 0 0 0 0 0 0 0 3 0 1 1 1 1 0 0 1
22 22 2 25 2 3 1 35 1 0 0 0 1 0 0 0 0 0 0 0 0 1 0 1 0 1 0 0 0 4 0 1 2 1 1 0 1 1
23 23 2 72 5 3 4 7 2 0 0 0 0 0 0 0 0 0 0 0 0 2 0 0 0 0 0 0 0 2 0 1 2 1 1 0 0 1
24 24 2 61 2 3 1 192 0 0 0 0 0 0 0 0 0 0 0 0 2 0 2 0 0 0 0 0 4 0 2 3 1 1 0 0 1
25 25 2 43 3 3 1 6 1 0 0 0 0 1 0 1 2 1 0 0 1 0 1 1 2 0 0 0 0 10 0 2 3 1 1 0 0 1
26 26 2 52 2 2 5 192 1 2 1 0 1 0 0 0 0 0 0 1 1 0 3 2 0 0 0 0 12 0 1 3 1 1 0 0 0
27 27 2 23 2 3 5 13 1 0 0 0 0 0 0 0 3 0 0 0 0 1 1 0 0 0 1 0 0 6 0 2 2 1 2 0 1 0
28 28 1 73 4 2 4 5 2 0 1 2 0 2 2 0 0 0 2 0 0 0 0 0 0 0 0 0 0 9 0 1 3 1 1 0 0 1
29 29 2 34 2 3 2 19 2 0 2 2 0 1 0 2 1 1 1 1 2 3 2 3 3 3 2 0 0 28 1 1 2 1 2 0 0 0
30 30 2 34 2 3 1 20 1 0 0 0 0 0 0 0 0 0 0 1 0 0 0 0 0 0 0 0 1 0 1 2 1 2 0 0 1
31 31 1 47 5 5 1 19 4 0 1 0 0 0 0 0 0 0 0 0 3 1 0 1 1 0 0 0 7 0 1 2 1 1 0 1 1
32 32 2 31 2 4 1 45 4 1 1 0 0 1 0 0 0 0 0 0 1 1 1 1 1 1 1 1 0 0 11 0 1 2 2 1 1 1 0
33 33 1 60 2 3 1 35 1 0 0 0 0 1 0 0 3 0 0 0 1 0 0 1 1 1 0 0 0 8 0 1 2 1 1 1 1 1
34 34 2 35 2 3 5 23 1 0 1 0 0 0 0 0 1 1 1 0 1 0 1 0 1 1 0 1 1 10 0 2 1 1 2 0 0 0
35 35 2 56 2 3 2 23 1 0 0 0 0 0 0 0 0 0 0 0 0 0 0 0 0 1 0 0 0 0 1 0 1 1 1 2 0 0 0
36 36 1 40 2 3 1 15 1 0 1 0 0 1 0 0 0 0 0 0 0 0 1 0 1 0 0 0 4 0 1 1 1 2 0 0 0
37 37 2 33 2 4 1 194 0 0 1 0 0 0 0 0 0 0 0 1 0 0 1 1 0 0 1 1 6 0 1 2 1 1 0 1 1
38 38 2 35 2 2 1 15 2 1 1 0 0 1 0 0 0 2 1 1 1 1 3 0 1 3 1 1 0 18 1 1 3 1 1 1 0 1
39 39 1 59 2 2 1 23 4 0 0 0 0 0 0 0 0 0 0 0 0 0 0 0 0 0 0 0 0 1 2 1 2 0 0 0
40 40 1 42 3 5 1 23 4 1 0 0 0 1 0 0 0 0 0 0 0 0 0 0 0 0 0 0 2 0 1 1 1 2 0 0 0
41 41 1 19 1 3 1 11 4 0 0 0 0 0 0 0 2 0 0 0 1 1 0 1 1 1 0 0 7 0 1 1 2 2 0 0 0
42 42 1 32 2 7 1 23 4 0 0 0 0 0 0 1 0 0 0 0 0 1 2 0 1 0 5 0 1 2 1 2 0 0 0
43 43 2 47 2 6 1 55 1 0 0 0 0 0 0 2 0 0 3 3 0 1 0 0 0 0 9 0 1 1 1 1 1 1 0
44 44 1 51 2 4 2 28 1 0 0 0 0 0 3 0 0 0 1 1 3 1 0 0 0 0 9 0 1 3 1 1 0 0 1
45 45 2 66 2 3 4 23 4 0 0 0 0 0 0 0 0 0 0 0 0 0 0 0 0 0 0 0 0 1 1 1 2 0 0 1
46 46 2 53 5 2 19 1 0 0 0 1 0 0 0 0 0 0 0 0 0 1 0 0 0 0 2 0 1 2 2 2 0 1 1
47 47 2 32 2 3 2 35 1 2 2 2 3 2 1 1 0 2 1 1 1 0 0 1 1 1 1 1 2 4 1 1 1 1 1 0 1 0
48 48 1 55 2 5 1 35 1 0 1 0 0 0 0 1 0 1 1 1 0 0 1 0 2 2 2 0 1 13 0 1 2 1 2 0 0 0
49 49 2 56 2 6 1 55 4 0 1 0 0 0 0 0 0 3 2 1 1 0 1 0 0 1 1 1 0 12 0 1 1 1 1 0 0 1
```

TABLE B.2

Variable Number	Variable Name in SAS Program	Description
1	obs	Observation
2	id	Identification number
3	gender	Gender: 1 = male; 2 = female
4	age	Age in years at last birthday
5	marital	Marital status: 1 = never married; 2 = married; 3 = divorced; 4 = separated; 5 = widowed
6	education	Level of education: 1 = less than high school; 2 = some high school; 3 = finished high school; 4 = some college; 5 = finished bachelor degree; 6 = finished master degree; 7 = finished doctorate
7	employment	Employment status: 1 = full time; 2 = part time; 3 = unemployed; 4 = retired; 5 = houseperson; 6 = in school; and 7 = other
8	income	Income in thousands of dollars per year
9	religion	Religion: 1 = Protestant; 2 = Catholic; 3 = Jewish; 4 = none; and 5 = other
10–29	statement1 – statement20	"Please look at this card and tell me the number that best describes how often you felt or behaved in this way in the past week." — 20 items from the depression scale: 0 = rarely or none of the time (less than 1 day); 1 = some or a little of the time (1–2 days); 2 = occasionally or a moderate amount of time (3–4 days); and 3 = most or all of the time (5–7 days)

Table B.2 (continued)

Variable Number	Variable Name in SAS Program	Description
30	cesd	Sum of Statement1 to Statement 20: 0 = lowest level possible; 60 = highest level possible
31	cases	0 = normal; 1 = depressed, where depressed is cesd ≥ 16
32	drink	Regular drinker?: 1 = yes; 2 = no
33	health	General health?: 1 = excellent; 2 = good; 3 = fair; and 4 = poor
34	regdoctor	Have a regular doctor?" 1 = yes; 2 = no
35	treat	"Has a doctor prescribed or recommended that you take medicine; medical treatments; or change your way of living in such areas as smoking, special diet, exercise or drinking?": 1 = yes; 2 = no
36	beddays	Spent entire day(s) in bed in the last two months?: 0 = no; 1 = yes
37	acuteill	Any acute illness in the last two months?: 0 = no; 1 = yes
38	chronicill	Any chronic illness in last two months?: 0 = no; 1 = yes

Source: Reprinted with permission from Afifi, A.A. and Clark, V., *Computer-Aided Multivariate Analysis*, 3rd ed., Chapman & Hall, London, 1996. Copyright CRC Press, Boca Raton, Florida.

PROGRAM B.1

```
/*
Author: Neil Spencer
Creation Date: ??/??/??
Revision 1 Date: ??/??/??
Purpose: Read Los Angeles Depression Data
*/
LIBNAME abcxyz 'a:\data';
LIBNAME library 'a:\formats';
PROC FORMAT LIBRARY = library.formats;
VALUE gendlab 1 = 'male' 2 = 'female';
VALUE maritlab 1 = 'never married'
  2 = 'married' 3 = 'divorced' 4 =
  'separated' 5 = 'widowed';
VALUE educlab 1 = 'less than high school'
  2 = 'some high school' 3 = 'finished
  high  school' 4 = 'some college' 5 =
  'finished bachelor degree' 6 = 'finished
  master degree' 7 = 'finished doctorate';
VALUE empllab 1 = 'full time' 2 = 'part
  time' 3 = 'unemployed' 4 = 'retired' 5
  = 'houseperson' 6 = 'in school' 7 =
  'other';
VALUE religlab 1 = 'Protestant' 2 =
  'Catholic' 3 = 'Jewish' 4 = 'none' 5 =
  'other';
VALUE stmntlab 0 = 'rarely or none of the
  time (less than day)' 1 = 'some or a
  little of the time (1-2 days)' 2 =
  'occasionally or a moderate amount of
  the time (3-4 days)' 3 = 'most or all
  of the time (5-7 days)';
VALUE caseslab 0 = 'normal' 1 =
  'depressed';
```

```
VALUE ynlaba 1 = 'yes' 2 = 'no';
VALUE hlthlab 1 = 'excellent' 2 = 'good'
  3 = 'fair' 4 = 'poor';
VALUE ynlabb 0 = 'no' 1 = 'yes';

DATA abcxyz.depress;

INFILE 'a:\data\depress.dat';
INPUT obs id gender age marital education
  employment income religion
  statement1-statement20 cesd cases
  drink health regdoctor treat beddays
  acuteill chronicill;
LABEL obs = 'Observation';
LABEL id = 'Indentification number';
LABEL gender = 'Gender';
LABEL age = 'Age';
LABEL marital = 'Marital status';
LABEL education = 'Level of education';
LABEL employment = 'Employment status';
LABEL income = 'Income';
LABEL religion = 'Religion';
LABEL statement1 = 'I felt that I could
not shake off the blues even with the help
of my family or friends';
LABEL statement2 = 'I felt depressed';
LABEL statement3 = 'I felt lonely';
LABEL statement4 = 'I had crying spells';
LABEL statement5 = 'I felt sad';
LABEL statement6 = 'I felt fearful';
LABEL statement7 = 'I thought my life had
been a failure';
LABEL statement8 = 'I felt that I was as
good as other people';
```

```
LABEL statement9 = 'I felt hopeful about
the future';
LABEL statement10 = 'I was happy';
LABEL statement11 = 'I enjoyed life';
LABEL statement12 = 'I was bothered by
things that usually do not bother me';
LABEL statement13 = 'I did not feel like
eating; my appetite was poor';
LABEL statement14 = 'I felt that everything
was an effort';
LABEL statement15 = 'My sleep was
restless';
LABEL statement16 = 'I could not get
going';
LABEL statement17 = 'I had trouble keeping
my mind on what I was doing';
LABEL statement18 = 'I talked less than
usual';
LABEL statement19 = 'People were
unfriendly';
LABEL statement20 = 'I felt that people
disliked me';
LABEL cesd = 'CESD score';
LABEL cases = 'Cases';
LABEL drink = 'Regular drinker';
LABEL health = 'General health';
LABEL regdoctor = 'Have a regular doctor';
LABEL treat = 'Treatment recommended';
LABEL beddays = 'Spent entire day(s) in
bed in last two months';
LABEL acuteill = 'Any acute illness in the
last two months?';
LABEL chronicill = 'Any chronic illness in
the last two months?';
```

```
FORMAT gender gendlab.;
FORMAT marital maritlab.;
FORMAT education educlab.;
FORMAT employment empllab.;
FORMAT religion religlab.;
FORMAT statement1-statement20 stmntlab.;
FORMAT cases caseslab.;
FORMAT drink regdoctor treat ynlaba.;
FORMAT health hlthlab.;
FORMAT beddays acuteill chronicill ynlabb.;
PROC PRINT data = abcxyz.depress;
RUN;
```

Appendix C

Solution Programs for Tasks

The programs below will produce the results requested in the tasks. However, deviations from the programming shown here may also produce equally acceptable results, particularly in relation to the presentation of output.

Task 2.1

```
/*
Author: Neil Spencer
Purpose: Task 2.1
*/
PROC FORMAT;
VALUE perlab 1 = 'Before program'
   2 = '1 month after' 3 = '2 months after'
   4 = '5 months after';
DATA GLOVES;
INFILE 'a:\data\gloves.dat' TRUNCOVER;
INPUT nurse 1-2 period 3 observed 4-5
   gloveuse 6-7 experience 8-9;
LABEL nurse = 'Nurse';
LABEL period = 'Observation period';
LABEL observed = 'Number of observations';
LABEL gloveuse = 'Number of times gloves
   used';
```

```
LABEL experience = 'Years experience';
FORMAT period perlab.;
prop = gloveuse/observed;
LABEL prop = 'Proportion of times gloves
  used';

PROC PRINT DATA = gloves;

RUN;
```

Task 2.2

Experimenting with programming.

Task 2.3

Experimenting with programming.

Task 2.4

```
/*
Author: Neil Spencer
Purpose: Task 2.4
*/

PROC IMPORT DATAFILE =
  'a:\data\glovesdelim.dat'
  OUT = gloves DBMS = DLM REPLACE;
DELIMITER = ';';
GETNAMES = YES;

PROC PRINT DATA = gloves;

RUN;
```

Task 3.1

```
/*
Author: Neil Spencer
```

```
Purpose: Task 3.1
*/
LIBNAME abcxyz 'a:\data';
LIBNAME library 'a:\formats';
PROC FORMAT LIBRARY = library.formats;
VALUE perlab 1 = 'Before program'
  2 = '1 month after' 3 = '2 months after'
  4 = '5 months after';
DATA abcxyz.gloves;
INFILE 'a:\data\gloves.dat' TRUNCOVER;
INPUT nurse 1-2 period 3 observed 4-5
  gloveuse 6-7 experience 8-9;
LABEL nurse = 'Nurse';
LABEL period = 'Observation period';
LABEL observed = 'Number of observations';
LABEL gloveuse = 'Number of times gloves
  used';
LABEL experience = 'Years experience';
FORMAT period perlab.;
prop = gloveuse/observed;
LABEL prop = 'Proportion of times gloves
  used';

RUN;
```

TASK 3.2

```
/*
Author: Neil Spencer
Purpose: Task 3.2
*/
LIBNAME abcxyz 'a:\data';
LIBNAME library 'a:\formats';
PROC CONTENTS DATA = abcxyz.gloves;
```

```
RUN;
```

TASK 3.3

```
/*
Author: Neil Spencer
Purpose: Task 3.3
*/

LIBNAME abcxyz 'a:\data';
LIBNAME library 'a:\formats';

PROC REPORT DATA = abcxyz.gloves;
COLUMN nurse period observed gloveuse prop
  experience;
TITLE 'Glove Use Data';
DEFINE nurse/ORDER WIDTH = 6 'Nurse Number'
  F = 2.0;
DEFINE period/ORDER WIDTH = 6 'Period'
  F = 1.0;
DEFINE observed/WIDTH = 6 'Number of times
  observed' F = 2.0;
DEFINE gloveuse/WIDTH = 6 'Number of glove
  uses' F = 2.0;
DEFINE prop/WIDTH = 12 'Proportion of times
  gloves used' F = 5.3;
DEFINE experience/WIDTH = 6 'Years
  experience' F = 2.0;

RUN;
```

TASK 3.4

```
/*
Author: Neil Spencer
Purpose: Task 3.4
*/

LIBNAME abcxyz 'a:\data';
```

```
LIBNAME library 'a:\formats';

DATA gloves;
SET abcxyz.gloves;
FILE 'a:\glovesput.dat';
PUT nurse 1-2 period 4 observed 6-7
 gloveuse9-10 experience 12-13
 prop 15-18.2;

RUN;
```

TASK 3.5
```
/*
Author: Neil Spencer
Purpose: Task 3.5
*/

LIBNAME abcxyz 'a:\data';
LIBNAME library 'a:\formats';

PROC EXPORT DATA = abcxyz.gloves DBMS = DLM
 OUTFILE = 'a:\glovesdelim2.dat' REPLACE;
DELIMITER = '^';

RUN;
```

TASK 4.1
```
/*
Author: Neil Spencer
Purpose: Task 4.1
*/

LIBNAME library 'a:\formats';

DATA glovesobs;
INFILE 'a:\data\glovesobs.dat' TRUNCOVER;
INPUT nurse 1-2 period 3 observed 4-5
   gloveuse 6-7;
LABEL nurse = 'Nurse';
```

```
LABEL period = 'Observation period';
LABEL observed = 'Number of observations';
LABEL gloveuse = 'Number of times gloves
  used';
FORMAT period perlab.;

DATA glovesexp;
INFILE 'a:\data\glovesexp.dat' TRUNCOVER;
INPUT nurse 1-2 experience 3-4;
LABEL nurse = 'Nurse';
LABEL experience = 'Years experience';

PROC SORT DATA = glovesobs;
BY nurse;

PROC SORT DATA = glovesexp;
BY nurse;

DATA gloves;
MERGE glovesobs glovesexp;
BY nurse;

PROC PRINT DATA = gloves;

RUN;
```

Task 4.2

Experimenting with programming.

Task 4.3

```
/*
Author: Neil Spencer
Purpose: Task 4.3
*/

LIBNAME abcxyz 'a:\data';
LIBNAME library 'a:\formats';

PROC FORMAT LIBRARY = library.formats;
```

```
VALUE grouplab 1 = 'Less than 8 years
  experience'2 = '8 years or more
experience';
DATA gloves;
SET abcxyz.gloves;
IF experience > = 8 THEN group = 2;
 ELSE group = 1;
LABEL group = 'Experience group';
FORMAT group grouplab.;
PROC PRINT DATA = gloves;

RUN;
```

Task 5.1
```
/*
Author: Neil Spencer
Purpose: Task 5.1
*/
LIBNAME abcxyz 'a:\data';
LIBNAME library 'a:\formats';

DATA glovesonerow;
INFILE 'a:\data\glovesonerow.dat'
TRUNCOVER;
INPUT nurse 1-5 obs1 6-9 obs2 10-13
  obs3 14-17 obs4 18-21 use1 22-25 use2
  26-29 use3 30-33 use4 34-37 experience
  38-41;
DATA abcxyz.glovessplit;
SET glovesonerow;
ARRAY obs[4] obs1-obs4;
ARRAY use[4] use1-use4;
DO I = 1 TO 4;
 period = i;
 observed = obs[i];
```

```
  gloveuse = use[i];
  OUTPUT;
END;

DROP obs1-obs4 use1-use4 i;

PROC PRINT DATA = abcxyz.glovessplit;

RUN;
```

TASK 5.2

```
/*
Author: Neil Spencer
Purpose: Task 5.2
*/

LIBNAME abcxyz 'a:\data';
LIBNAME library 'a:\formats';

DATA gloves;
SET abcxyz.glovessplit;

PROC SORT DATA = gloves;
BY nurse period;

DATA glovesonerow;
SET gloves;
BY nurse period;

ARRAY obs[4] obs1-obs4;
ARRAY use[4] use1-use4;
RETAIN obs1-obs4 use1-use4;
obs[period] = observed;
use[period] = gloveuse;
IF LAST.nurse THEN OUTPUT;

DROP period observed gloveuse;

PROC PRINT DATA = glovesonerow;

RUN;
```

TASK 6.1

```
/*
Author: Neil Spencer
Purpose: Task 6.1
*/

LIBNAME abcxyz 'a:\data';
LIBNAME library 'a:\formats';

PROC MEANS DATA = abcxyz.gloves MEAN CLM
  MIN MEDIAN MAX;
VAR prop;

RUN;
```

TASK 6.2

```
/*
Author: Neil Spencer
Purpose: Task 6.2
*/

LIBNAME abcxyz 'a:\data';
LIBNAME library 'a:\formats';

PROC UNIVARIATE DATA = abcxyz.gloves;
VAR prop;
PROBPLOT prop/NORMAL;

RUN;
```

TASK 6.3

```
/*
Author: Neil Spencer
Purpose: Task 6.3
*/

LIBNAME abcxyz 'a:\data';
LIBNAME library 'a:\formats';
```

```
PROC FREQ DATA = abcxyz.gloves;
TABLES period;

RUN;
```

TASK 6.4

```
/*
Author: Neil Spencer
Purpose: Task 6.4
*/

LIBNAME abcxyz 'a:\data';
LIBNAME library 'a:\formats';

PROC FREQ DATA = abcxyz.gloves;
TABLES nurse*period;

RUN;
```

TASK 7.1

```
/*
Author: Neil Spencer
Purpose: Task 7.1
*/

LIBNAME abcxyz 'a:\data';
LIBNAME library 'a:\formats';

PROC MEANS DATA = abcxyz.gloves MEAN CLM
  MIN MEDIAN MAX MAXDEC = 2;
VAR prop;
CLASS period;

RUN;
```

TASK 7.2

```
/*
Author: Neil Spencer
```

```
Purpose: Task 7.2
*/
LIBNAME abcxyz 'a:\data';
LIBNAME library 'a:\formats';

PROC TABULATE DATA = abcxyz.gloves;
CLASS period;
VAR prop;
TABLE period*prop,MEAN MIN MEDIAN MAX;

RUN;
```

TASK 7.3

```
/*
Author: Neil Spencer
Purpose: Task 7.3
*/
LIBNAME abcxyz 'a:\data';
LIBNAME library 'a:\formats';

PROC TABULATE DATA = abcxyz.gloves;
CLASS nurse period;
TABLE nurse*period;

RUN;
```

TASK 8.1

Choice of programming.

TASK 8.2

```
/*
Author: Neil Spencer
Purpose: Task 8.2
*/
LIBNAME abcxyz 'a:\data';
```

```
LIBNAME library 'a:\formats';
PROC FREQ DATA = abcxyz.gloves;
TABLES nurse*period/OUT = tableoutput;
PROC PRINT DATA = tableoutput;
RUN;
```

Task 9.1

```
/*
Author: Neil Spencer
Purpose: Task9.1
*/
LIBNAME abcxyz 'a:\data';
LIBNAME library 'a:\formats';
PROC UNIVARIATE DATA = abcxyz.depress
  NOPRINT;
VAR income;
HISTOGRAM income/ENDPOINTS = 0 TO 60 BY 5
  VSCALE = COUNT VAXISLABEL = 'Cases'
  NORMAL(NOPRINT);
RUN;
```

Task 9.2

```
/*
Author: Neil Spencer
Purpose: Task9.2
*/
LIBNAME abcxyz 'a:\data';
LIBNAME library 'a:\formats';
PROC GCHART DATA = abcxyz.depress;
HBAR education/DISCRETE SUMVAR = income
  TYPE = MEAN;
```

```
PATTERN  COLOR = BLUE  VALUE = L3;
RUN;
```

TASK 9.3

```
/*
Author: Neil Spencer
Purpose: Task9.3
*/
LIBNAME abcxyz 'a:\data';
LIBNAME library 'a:\formats';
PROC GCHART DATA = abcxyz.depress;
HBAR education/DISCRETE SUBGROUP =
  employment NOSTATS;
PATTERN1  COLOR = GREEN  VALUE = L2;
PATTERN2  COLOR = RED  VALUE = L2;
PATTERN3  COLOR = BLUE  VALUE = L2;
PATTERN4  COLOR = BLACK  VALUE = L2;
PATTERN5  COLOR = GRAY  VALUE = L2;
PATTERN6  COLOR = CYAN  VALUE = L2;
PATTERN7  COLOR = MAGENTA  VALUE = L2;

RUN;
```

TASK 9.4

```
/*
Author: Neil Spencer
Purpose: Task9.4
*/
LIBNAME abcxyz 'a:\data';
LIBNAME library 'a:\formats';
PROC GCHART DATA = abcxyz.depress;
PIE education/DISCRETE VALUE = ARROW
  TYPE = FREQ ANGLE = 190;
```

```
PATTERN1  COLOR = GREEN VALUE = P1N45;
PATTERN2  COLOR = RED VALUE = P1N45;
PATTERN3  COLOR = BLUE VALUE = P1N45;
PATTERN4  COLOR = BLACK VALUE = P1N45;
PATTERN5  COLOR = GRAY VALUE = P1N45;
PATTERN6  COLOR = CYAN VALUE = P1N45;
PATTERN7  COLOR = MAGENTA VALUE = P1N45;

RUN;
```

TASK 10.1

```
/*
Author: Neil Spencer
Purpose: Task 10.1
*/

OPTIONS MPRINT SYMBOLGEN;

DATA glovesonerow;
INFILE 'a:\data\glovesonerow.dat'
  TRUNCOVER;
INPUT nurse 1-5 obs1 6-9 obs2 10-13
  obs3 14-17 obs4 18-21 use1 22-25
  use2 26-29 use3 30-33 use4
  34-37 experience 38-41;
%INCLUDE 'a:\task10_1macro.sas';
%freqmac(obs1,use1)
%freqmac(obs2,use2)
%freqmac(obs3,use3)
%freqmac(obs4,use4)

RUN;
```

TASK 10.1 MACRO

```
/*
Author: Neil Spencer
Purpose: Task 10.1 macro
```

```
*/
%MACRO freqmac(var1,var2);
PROC FREQ DATA = glovesonerow;
TABLE &var1*&var2;
%MEND;
```

TASK 10.2

```
/*
Author: Neil Spencer
Purpose: Task 10.2
*/
OPTIONS MPRINT SYMBOLGEN;
LIBNAME abcxyz 'a:\data';
LIBNAME library 'a:\formats';
%INCLUDE 'a:\task10_2macro.sas';
%barchart(statement5)
%barchart(statement10)
%barchart(statement11)

RUN;
```

TASK 10.2 MACRO

```
/*
Author: Neil Spencer
Purpose: Task 10.2 macro
*/
%MACRO barchart(var1);
PROC GCHART DATA = abcxyz.depress;
HBAR &var1/DISCRETE NOSTATS;
%MEND;
```

Index